Jörg Bandura

From cell to robot

Jörg Bandura

From cell to robot

A bio-inspired locomotion device

Südwestdeutscher Verlag für Hochschulschriften

Impressum / Imprint
Bibliografische Information der Deutschen Nationalbibliothek: Die Deutsche Nationalbibliothek verzeichnet diese Publikation in der Deutschen Nationalbibliografie; detaillierte bibliografische Daten sind im Internet über http://dnb.d-nb.de abrufbar.
Alle in diesem Buch genannten Marken und Produktnamen unterliegen warenzeichen-, marken- oder patentrechtlichem Schutz bzw. sind Warenzeichen oder eingetragene Warenzeichen der jeweiligen Inhaber. Die Wiedergabe von Marken, Produktnamen, Gebrauchsnamen, Handelsnamen, Warenbezeichnungen u.s.w. in diesem Werk berechtigt auch ohne besondere Kennzeichnung nicht zu der Annahme, dass solche Namen im Sinne der Warenzeichen- und Markenschutzgesetzgebung als frei zu betrachten wären und daher von jedermann benutzt werden dürften.

Bibliographic information published by the Deutsche Nationalbibliothek: The Deutsche Nationalbibliothek lists this publication in the Deutsche Nationalbibliografie; detailed bibliographic data are available in the Internet at http://dnb.d-nb.de.
Any brand names and product names mentioned in this book are subject to trademark, brand or patent protection and are trademarks or registered trademarks of their respective holders. The use of brand names, product names, common names, trade names, product descriptions etc. even without a particular marking in this work is in no way to be construed to mean that such names may be regarded as unrestricted in respect of trademark and brand protection legislation and could thus be used by anyone.

Coverbild / Cover image: www.ingimage.com

Verlag / Publisher:
Südwestdeutscher Verlag für Hochschulschriften
ist ein Imprint der / is a trademark of
OmniScriptum GmbH & Co. KG
Heinrich-Böcking-Str. 6-8, 66121 Saarbrücken, Deutschland / Germany
Email: info@svh-verlag.de

Herstellung: siehe letzte Seite /
Printed at: see last page
ISBN: 978-3-8381-5156-4

Zugl. / Approved by: Bonn, Rheinische-Friedrich-Wilhelms-Universität, Dissertation, 2014

Copyright © 2015 OmniScriptum GmbH & Co. KG
Alle Rechte vorbehalten. / All rights reserved. Saarbrücken 2015

Summary Bionics or biomimetics is an interdisciplinary research field, a scientific approach to applicate naturally developed biological systems, methods and solutions to the study and design of technology and engineering systems. Therefore bionics is based on an exclusive mutuality between life sciences and technology and its associated sciences, such as robotics. Robots are special artificial agents, and they have much in common with biological agents in case of the need to adapt to their environment. A popular trend in robotics is the development of soft robots – artificial agents with a rather flexible skin or shape, propulsing itself with some type of crawling movement. These robots are able to deform and adapt to obstacles during locomotion, which is an advantage over classical wheeled or legged propulsion. Bionics is helpful in developing locomotion devices for robots, e. g. bio-inspired climbing robots, such as geckobots, utilise the biological gecko adhesion model for climbing. Most of these bio-inspired climbing robots have the disadvantage of using legs for locomotion. The idea is to find a new biological model for a bionic robotic locomotion device that is using an adhesion-dependent crawling locomotion, which allows the robot to climb (or at least be able to master inclinations) and still has a rather soft and deformable shape providing the flexibility of adaptation to obstacles or a changing environment. Surprisingly, single cells, such as amoebae or animal tissue cells, provide these required properties: the ability to crawl on surfaces by formation of adhesion bonds and a very deformable shape – a perfect model for such robots. These cells are reorganising their cytoskeletal cortex and create a visco-elastic gradient which is polarising the cell with a sol-like "sloppy" leading edge at the front and a gel-like "stiff" rear end. This work demonstrates that it is possible to transfer the biophysical locomotion mechanism of cell migration to a simulation model of soft robots, which use an adhesion-dependent mechanism to autonomously create a polarising elasticity gradient during motion. It introduces and analyses three robot models, which are able to move on surfaces with different built-in integrations of this polarisation mechanism. Simulations show that the robots are flexible enough to adapt to changing environments, such as rough surfaces. One model is even able to crawl on walls and ceilings against the direction of gravity. Finally, this work offers some ideas for possible constructions and usability of these robots, and what insights their analysis might give into principles of biological cell migration.

"Let us consider what bionics has come to mean operationally and what it or some word like it (I prefer biomimetics) ought to mean in order to make good use of the technical skills of scientists specializing, or rather, I should say, despecializing into this area of research. Presumably our common interest is in examining biological phenomenology in the hope of gaining insight and inspiration for developing physical or composite biophysical systems in the image of life."

—Otto Herbert Schmitt, 1963

Contents

About Bionics	1
1 Introduction	7
1.1 Robotic locomotion	7
1.2 Biological cell locomotion	9
1.2.1 Biophysics of cell migration	11
1.2.2 Bionic abstraction	20
2 Modelling	23
2.1 Introduction of robot models	23
2.1.1 Configuration and notation	26
2.1.2 Non-dimensionalisation	26
2.2 Model mechanics	27
2.2.1 Static forces	27
2.2.2 Friction forces	29
2.2.3 Dynamics	32
2.3 Surface roughness	34
2.3.1 Modelling	35
2.3.2 Surface adaptation	37
2.4 Parameter overview	37
3 Simulation results	41
3.1 Overall performance	41
3.1.1 Translocation speed	43
3.1.2 Adhesiveness	45
3.1.3 Polarity	45
3.1.4 Forces	49
3.1.5 Mechanical stress	55
3.1.6 Correlations	57
3.2 Parameter screening	57
3.2.1 Elasticity	59
3.2.2 Bending stiffness	61
3.2.3 Elasticity adaptation	62
3.3 Rough surface performance	63
3.3.1 Translocation speed	63
3.3.2 Adhesiveness	64
3.4 Capabilities	66
4 Concluding evaluation	69
4.1 Constructability	70
4.2 Usability	72
4.3 Reverse bionics	73
4.4 Outlook	74
A Appendix	75
A.1 Mathematical derivation of friction models	75
A.2 Smart material actuators	79
B Supplementary material	85
References	87

List of figures

0.1	Lilienthal: Our instructors of flight	5
1.1	Deformable soft robot	8
1.2	"Deformable Wheel" robot	8
1.3	Cell migration schematics	13
1.4	Cell abstraction	21
2.1	The three models	24
2.2	Configuration of vertices, segments and vectors	26
2.3	Torsion spring response	29
2.4	Surface roughness: dx	36
2.5	Surface roughness: σ	36
2.6	Surface roughness: L_c	37
2.7	Surface profiles overview	39
3.1	Persistent movement	42
3.2	Translocation speed	44
3.3	Adhesiveness	46
3.4	Chain curvature	47
3.5	Polarity	48
3.6	Forces in model	50
3.7	Disruption forces	51
3.8	Traction forces	53
3.9	Force course	54
3.10	Mechanical stress	56
3.11	Adhesiveness correlations	58
3.12	Polarity correlations	59
3.13	Parameter screening: k_e	60
3.14	Parameter screening: k_m	61
3.15	Parameter screening: r	62
3.16	Roughness: locomotion	64
3.17	Roughness: speed	65
3.18	Roughness: height	65
3.19	Tube movement	66
3.20	Angular speed	67
3.21	Angular polarity	67
3.22	Angular correlations	68
4.1	Jamming skin enabled locomotion (JSEL) schematics	72
4.2	JSEL robot	72

About Bionics

BIONICS – the undiscovered country. The term *bionics* is usually defined as a portmanteau from *biology* and *technics* and is describing the scientific approach to applicate naturally developed biological systems, methods and solutions to the study and design of technology and engineering systems. It is an interesting concept, because biological and technical systems have to cope with similar or same problems and need to work within the same limits given by the same physical conditions of this world. Additionally, both biological and technical systems share many similarities. Both are typically constructed systems, build of many small parts, which are enhanced by synergetic effects when combined. The combination results in a new quality: a function. This supplemental functional component is the main property of every biological and technical system.

The biological system is enhanced and maintained by an evolutionary process, which not only brought a manifold diversity of different forms of life, but which is also adapting and optimising life by these evolutionary principles on a time scale of millions of years. Hence the evolution of life has no predefined goal except this optimisation and adaptation to current prevailing environmental conditions. Evolution is a stochastic process, every genotype and every phenotype of a living organism is a variety of a set of inherited variables and parameters and therefore it has many coupled degrees of freedom. In a technical or mathematical sense it is comparable with a Monte Carlo simulation: repeated random sampling is converging to an optimum mean (according to the law of large numbers). Additionally, a beneficial mutation of an organism is enhancing the survivability and the evolutionary fitness of this organism, which significantly increases the probability that this beneficial mutation will prevail in future generations.

The technical system is an intelligent design. It is invented, planned, developed, built, enhanced and maintained by a human creator. There is always a plan and a target for each technical creation – it is planned and adapted for a certain purpose in advance. It begins with a prototype, which is getting improved, enhanced and advanced. In the case that it has proven to be a successful technology, it will be improved further for many generations until it might get replaced by a better and more efficient technology someday. The human mind is the perfect tool for this intelligent designing, because of its ability of creative thinking and abstracting. For millennia humanity has developed and improved many great and interesting machineries and technologies (ever since humans were able to use their minds in combination with their hands), though the human mind is only a limited resource. It has many degrees of freedom in thinking, but additionally, it often tends to be conservative, having predefined paths of thinking and does not leave them, if they have proven in functionality. Only the most genius minds are dare to sometimes leave the predefined paths and explore new ways leading to the fields of innovations.

Evolution and the biological system have none of these 'restrictions' – a stochastic process probes any probable possibility. This is why bionics is an interesting scientific approach. It is opening new paths of thinking, new ideas, new concepts and new solutions for technical problems. A new exercise for the human mind: looking at a naturally designed system, understanding and deducing the principle behind this system and then transferring it to innovative technology, which is achieving a similar purpose as the natural system (of course, this bio-inspired technology has to have an advantage compared to classical non-bio-inspired technology). The term *bio-inspiration* is nicely describing the aim and the concepts behind bionics. It is a common misconception, that bionics is just about copying nature and rebuilding it (or even replacing it) – quite in contrary, the aim of bionics is to understand the abstract principles behind a biological system and to use this newly gathered knowledge for transfer into technology. A bionic invention and technology normally does not look anyhow similar to its natural example.

The way of information processing of a bionic approach is either a top-down or bottom-up strategy. In the first case, there is a certain technical limitation or problem, which is compared to a natural system, by investigating how nature is handling similar or the same limitations and then adapting the natural system for handling the technical system. The bottom-up strategy is working the other way around – by observing and examining nature, different interesting structures and adaptations are revealed, which then might be transferred to enhance a current technology or even lead to the invention of a new technology. That is one reason, why biodiversity and pure research is very important for bionics, because the gained knowledge and the understanding of fundamental principles is essential for a possible transfer into technology. Even if pure biological research does not yield an immediate commercial benefit, it may become commercially interesting later by improving technology. Finally, bionics is bringing biologists and engineers together, forming a new cooperation between very different specialists – a clash of sciences leading to the birth of new ideas.

Another benefit of bionics is the reverse information processing way – 'reverse bionics'. Simulating, modelling and rebuilding natural solutions in technical applications helps to understand nature's structures and systems, to answer why nature is using these structures and systems and why they are designed this way, leading to a deeper understanding of natural processes beyond pure descriptive analysis and enhancing their functional analysis.

Terminology

The terms *bionics* and *biomimetics* are often used as synonyms:

Biomimetics (from Greek βίος "life" and μιμητικός "imitative") was coined by he American inventor, engineer and biophysicist Otto Herbert Schmitt (6th April 1913 – 6th January 1998) during the 1950s [114, 104]. He was known for his scientific research on biophysics and his focus on devices that mimic natural systems. He developed the Schmitt trigger by studying the nerves in squids and tried to engineer a device that technically replicated the biological nerve propagation system.

> "Biophysics is not so much a subject matter as it is a point of view. It is an approach to problems of biological science utilizing the theory and technology of the physical sciences. Conversely, biophysics is also a biologist's approach to problems of physical science and engineering, although this aspect has largely been neglected."
>
> —Otto Herbert Schmitt, 1957 [87]

Bionics was coined by Jack Ellwood Steele (27th January 1924 – 19th January 2009) in 1958, an American medical doctor and US Air Force colonel, working at the Aeronautics Division House at Wright-Patterson Air Force Base in Dayton, Ohio, USA [115, 120]. He studied biological organisms to find solutions to engineering problems – he defined bionics as "the science of systems which have some function copied from nature, or which represent characteristics of natural systems or their analogues". The term *bionics* was officially used as the title of a three-day symposium in September 1960 [113].

A decade later Steele's work on bionics and the US Air Force research on cyborgs became popular in common literature and television. The 1972 novel *Cyborg* by science fiction author Martin Caidin contains explicit references to Steele. The book formed the basis of the 70s American TV series *The Six Million Dollar Man* featuring Lee Majors as astronaut and test pilot Steve Austin, who was severely injured during an aircraft crash and whose body parts where replaced by "bionic" implants worth of six million US-Dollar (though comparing the economies in the 1970s and now, it would be more likely six billion or more nowadays). The TV series and its spin-off *The Bionic Woman* popularised, if somewhat inaccurately, the term *bionics* [115].

1974 the word *biomimetics* made its first appearance in Webster's Dictionary, defined as "the study of the formation, structure, or function of biologically produced substances and materials (as enzymes or silk) and biological mechanisms and processes (as protein synthesis or photosynthesis) especially for the purpose of synthesizing similar products by artificial mechanisms which mimic natural ones."

Early history

The concept of looking at nature as a model for technological inventions finally got a name in the 1950s, but its tradition is much older.

Architecture is a common field of bionics, because nature has developed many static light-weight and robust structures which are adaptable (not only) for buildings. In the late 1940s the American architect and engineer Richard Buckminster Fuller (17th July 1895 – 1st July 1983) built the geodesic domes, buildings with a polyhedral spherical or partial-spherical shell structure [116]. These domes are used worldwide as parts of military radar stations, civic buildings and attractions in theme parks. They have the advantage of high stability with only low requirement of materials – a systemic concept that Fuller saw in nature for economic efficiency in usage of material and energy. He expanded this to his technological concept of ephemeralization and to the usage of synergies (two terms he coined). Nonetheless Walther Bauersfeld (23rd January 1879 – 28th October 1959) built a similar dome-structure for the Zeiss Planetarium in Jena (Germany) some twenty years before [123], but nothing is known if this was a bio-inspired concept or just the analogous result of thinking about an efficient dome-like construction. Although Fuller was not the original inventor, he can be credited for the full intrinsic mathematics of a geodesic dome and its popularisation, as well as the systemic view on nature that is required for bionics.

One of the most famous bio-inspired inventions (that was originating before the 1950s) is known under the brand name Velcro, the hook–and–loop fastener everyone has definitely used in some case. It was invented by the Swiss electrical engineer George de Mestral (19th June 1907 – 8th February 1990), who lived in Commugny, Switzerland [118]. The first conceptualisation of Velcro began in 1941, after de Mestral returned with his dog from a hunting trip in the Alps. The burdock burrs that stuck to his clothes and the dog's fur caught his attention about their working mechanism. After microscopic examination he noticed dozens of small hooks which are able to catch anything with a loop, such as textile fabrics, animal fur and hairs, etc. De Mestral saw the possibility that this is a quite simple method to bind two materials reversibly. No one took him and his idea seriously, so that he needed to find the proper materials on his own. The development of a working mechanised manufacturing process took about ten years, when he submitted his idea for patent in 1951 which was finally granted in 1955.

Another bio-inspired technological application was built in the early 20th century, which became an essential part of modern transportation and logistics – the two American Wright brothers, Orville (19th August 1871 – 30th January 1948) and Wilbur (16th April 1867 – 30th May 1912), made the first controlled, powered and sustained heavier-than-air flight of a human on 17th December 1903 and invented and built the first successful prototype of an airplane [122]. This was made possible by the previous work of the German aviation pioneer Otto Lilienthal (23rd May 1848 – 10th August 1896). Lilienthal identified the physical principle of birdflight and the importance of wing shape, summarised in his famous book in 1889 [Figure 0.1] [57]. His self-developed and self-built hang gliders

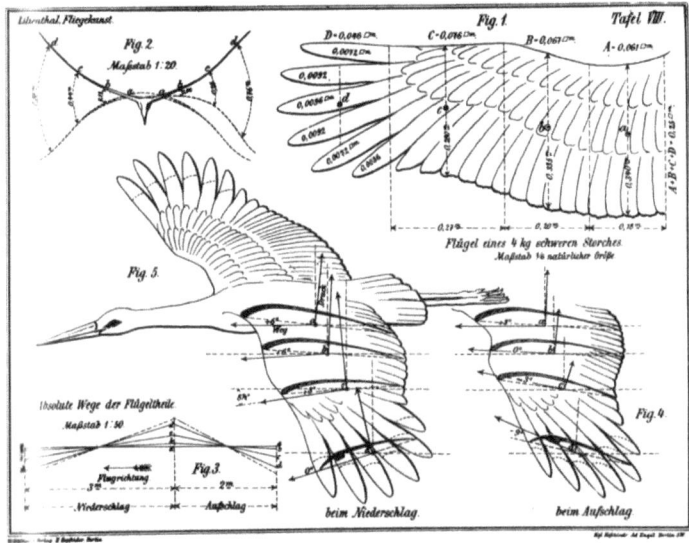

Figure 0.1.: "Unsere Lehrmeister im Fluge" (Our instructors of flight) – Otto Lilienthal's drawing of a White Stork (*Ciconia ciconia*) demonstrating the working principle of a bird's wing in his book *Der Vogelflug als Grundlage der Fliegekunst* (*Birdflight as the Basis of Aviation*), Berlin 1889 [57].

are the prototypes of human aviation, although he was killed in an accident with them. The Wright brothers were inspired by Lilienthal's work and they adapted, refined and enhanced his glider concepts, which finally led to "Kitty Hawk", the first motorised aircraft prototype in 1903 [122]. From this time on, every airplane is using the cross-sectional wing shape which Lilienthal had identified previously in a bird's wing.

Looking into earlier history, the Italian Renaissance artist, scientist and inventor Leonardo da Vinci (15th April 1452 – 2nd May 1519, st. v.) was one of the first, who was bio-inspired in the invention process. Like Lilienthal da Vinci was fascinated by flight – one famous work is his Codex on the Flight of Birds (1505) [117], which contains an examination of the flight behaviour of birds and the proposal of mechanisms for flight by machines. In the codex he noted for the first time that the center of gravity of a flying bird and its center of pressure do not coincide. Da Vinci constructed some of his proposed machines, and attempted to launch them from a hill near Florence, but, contrary to Lilienthal, his efforts failed. One reason of the failure might be that da Vinci attempted a more direct copy approach of nature, missing a higher degree of abstraction or missing the correct rescaling according to size, because of less available deeper knowledge of underlying physical principles.

Modern concepts and applications

Bionics is an interdisciplinary field of study, thus it is profiting from other interdisciplinary studies especially on the theoretical side, required by the bionic abstraction process. Two notable interdisciplinary theoretical studies that help to understand bionics are systems theory, which examines systems in general to elucidate their principles (applicable to all systems and research), and theoretical biology, which brings mathematics and biology closer together by providing appropriate theories and modelling tools. These are functioning as translators, helping in the communication between biologists, mathematicians, physicists and engineers, which often are using different specialised scientific languages. During the last decades biological surface structures and their properties became one central focus of bionic research due to the fact that these properties often are interesting for a technical application, providing superhydrophobicity, superhydrophilicity, friction reduction, friction enhancement, abrasion resistance, oil absorption, light harvesting, light reflection, under–water air retaining, shock absorption and much more. One famous bionic invention of the last decades is known as *lotus effect*[14], describing syperhydrophobic water repellent and self cleaning surfaces that were inspired by the leaf surface structure of *Nelumbo* plants, commonly known as lotus. Biological surfaces interesting for technological adaptations are not limited to plants, one animal example of interesting surface properties: the sandfish (*Scincus scincus*), a reptile species of skink living in desert regions in northern Africa, is able to burrow into the sand. The sandfish's scales have a surface with abrasive resistant properties, providing the sandfish with a low friction coefficient [15], allowing the sandfish to "swim" through sand. A technological adaptation might be a new surface coating with abrasive resistance for technological devices.

The significant increase of computation power since the 1950s helped to establish bio-inspired designs in (bio-)informatics. The concept of swarm intelligence in informatics and robotics is based on the swarm behaviour in nature, including ant colonies, bird flocking, animal herding, bacterial growth and fish schooling. Furthermore, artificial neural networks as a computational model of information processing are clearly inspired by function and structure of biological neural networks.

Besides, the neurological and senso-motoric systems found in nature are also utilised as a model for technical sensor and control systems. Examples are bio-inspired micro air vehicles, which do not only try to mimic the flapping wing mechanisms of insects for increased manoeuvrability but also need a simplified sensing and information processing system, that is also inspired by the simple insect's neuro-structure [112, 36]. Biological locomotion systems and locomotion itself is another central topic of bionic research. The exploration of biped, quadruped and six-legged propulsion for robots and machines are all inspired by nature. Even other types of natural locomotion systems are models for technical propulsion devices – e. g. fishes as models for aquatic locomotion, the snake's limbless crawling motion as model for terrestrial and/or aquatic propulsion [25].

1. Introduction

ROBOTICS – technology of artificial agents for certain autonomous applications, dealing with their design, construction, control and operation. The idea of the creation of autonomous machines dates back to classical times, though the research into their potential and functionality significantly increased in the 20th century due to the substantial increase in available technology required to build and operate such machines, currently making robotics a rapidly growing field of research [72].

The word robot comes from the Slavic word *robota*, used to refer to forced labour. The Czech writer Karel Čapek introduced the term in his play Rossum's Universal Robots, which premiered in 1921 [121].

According to the Oxford English Dictionary the term robotics was first used by American author and professor of biochemistry Isaac Asimov (2nd January 1920 – 6th April 1992) in his science fiction short story "Liar!", published in May 1941 in Astounding Science Fiction, but he was unaware of the coining because he assumed that robotics is already referring to the science and technology of robots, since electronics is referring to the science and technology of electrical devices [119].

1.1. Robotic locomotion

An important point of interest for robotics is locomotion that significantly enhances the usability compared to stationary robots. Classic robotic locomotion is using wheels because wheels have a long tradition in engineering as a simple to realise and easy to use locomotion device. However, their limitations are obvious, wheels are requiring a rather flat and even surface to operate properly.

To encounter the limitations of wheeled locomotion, research in robotics considers soft and/or deformable robots – agents that have a rolling movement similar to wheels or a crawling behaviour, but their (outer) shape is rather soft and/or deformable. The advantages compared to wheeled propulsion: these robots are able to adapt to rough terrain and inclinations and they are able to overcome obstacles in their path. Final concepts go one step further: the trial to construct a fully soft robot with the ability to squeeze itself through holes. One prototype of a deformable soft robot that is able to crawl and jump was introduced by Sugiyama and Hirai in 2006 [92]. They built a small circular wheel-like device with radial spokes and a deformable shape made of SMA coils as thermal actuators [Figure 1.1a]. For more information about SMAs see section A.2. By applying a

1. Introduction

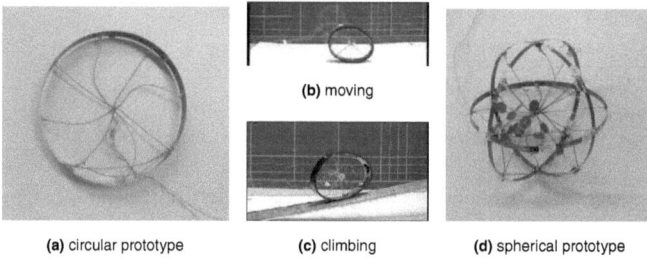

(a) circular prototype   (b) moving   (c) climbing   (d) spherical prototype

Figure 1.1.: The deformable soft robot constructed by Sugiyama and Hirai (2006) [92].

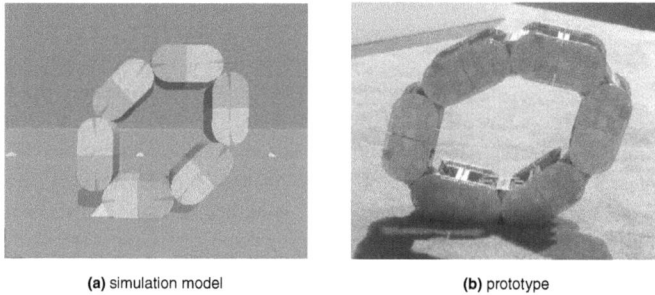

(a) simulation model   (b) prototype

Figure 1.2.: The "Deformable Wheel" robot as simulation model and constructed prototype by Chiu, Rubenstein and Shen (2008) [24].

current with a certain voltage pattern to each SMA coil, they are able to control its shape deformation and propel the device with a crawling/rolling like movement [Figure 1.1b]. The device has also the potential to manage inclinations [Figure 1.1c]. Besides, it is able to create high potential energy by shape-deformations, which it can use to jump. One step further they created a spherical deformable robot [Figure 1.1d], which has the same abilities as its two-dimensional counterpart [92].

Another interesting prototype was presented by Chiu, Rubenstein and Shen during a symposium in Japan in November 2008: the "Deformable Wheel" – a self-recovering modular rolling track [24]. This kind of robotic locomotion device has no soft shape as the previous introduced robots, but it is using a similar type of movement, a mixture of crawling and rolling. The Deformable Wheel has rigid segments, which are connected by joints to form a circular shape [Figure 1.2]. Its movement is regulated by the angles of the joints which are motorised by the segments. This robot has the ability to autonomously "stand up" after a knock over and to recover from a fall from heights.

However, these are not bionic or bio-inspired robots. Bio-inspired robotics for terrestrial locomotion is concentrating on biomechanics of legged or limbless locomotion on even surfaces and on crawling and especially climbing. These are interesting types of movement, because they are heavily dependent on (surface) friction and by enhancing or lowering this friction the propulsion can be adapted and adjusted to different surface properties and environments. Biological organisms often enhance friction by adhesion that is mostly based on molecular or atomic interactions. An example: a famous biological model in bionics is the gecko, which is able to "stick" to nearly every surface and on walls and ceilings alike, because the gecko toe's special microscale seta structure interacts via van der Waals forces with the surface structure on a molecular level [10].

Bio-inspired robots with a crawling/climbing movement can be divided into the following groups: *geckobots* [99] – using van der Waals forces (mimicking the Gecko) but working only on smooth surfaces; *stickybots* [53, 86, 9, 98] – using dry adhesives, working best on smooth surfaces; and *spinybots* [89] – insect-like hexapodal locomotion devices using spines. All of these robots share the common limitations of legged climbing: difficulties in handling obstacles (no flexibility) or the transition between horizontal and vertical direction and the inability to climb on both smooth and rough surfaces. Besides, legs are requiring wider space for their coordination. Thus, the bio-inspired crawling/climbing robots have some limitations, mainly because they focus on legged locomotion. The best idea would be to combine their abilities with the flexibility of the soft and deformable robots introduced earlier. Or, in other words, these small soft and deformable robots, which are able to crawl, just need to learn to climb. Bionic research is concentrating on adapting biological functions. Is there a life form on this planet that is small and flexible to deform itself, adapting and adhering to different surfaces which allows this life form to crawl on flat surfaces and climb on walls? The answer is simple: yes! This life form is one of the smallest of this planet: a cell.

Properties of a single cell (and the associated cellular mechanisms) are often neglected for bionic models (of course cells have a scaling advantage according to their small size, so some physical circumstances can be neglected or rather play a smaller role due to smaller size, which might not be to easily overcome on a larger scale on a technological level). Nonetheless this project is trying something new by modelling a locomotion device that is inspired by the biophysical properties and mechanics of cell migration, a type of propulsion commonly used by single tissue cells, which allows them to crawl on a surface. The next section is introducing the biology and the involved physics of this movement type.

1.2. Biological cell locomotion

Cellular movement is multifaceted and most cells are capable to propel themselves in some way, from small singular bacteria to eukaryotic cells embedded in multicellular organisms.

Bacteria are able to swim through water by using flagella, bacterial gliding and twitching motility allows them to move across surfaces, and the ability to control their buoyancy

1. Introduction

allow for vertical movement in water [13]. Flagella are semi-rigid protein polymer molecules forming a cylindrical whip-like structure that is rotated and used like a propeller for propulsion, motorised by a reversible motor at the base using an electrochemical gradient across the membrane for power [60]. Due to the bacterial size advantage, they are able to operate at a low Reynolds number, allowing them to swim with a relatively fast speed between 10 to 100 cell length per second [33]. In twitching motility, bacteria are using their pili as some kind of grappling hook: they can repeatedly extend it, anchoring it and then retracting it with considerable force [63].

Eukaryotic cells have undulipodia – cilia and flagella – available for swimming, though the difference to the bacterial flagella is that these are tubular membrane extensions rather than external protein polymer structures. The cross-section of one undulipodium is characterised by the "9+2" structure: nine fused pairs of microtubules, crosslinked to each neighbour are circularly surrounding two single microtubules in the centre, this structure is forming the so-called axoneme [45].

However, many eukaryotic cells crawl across a surface rather than swim with cilia or flagella. It is observable from singular predatory amoebae [5] to tissue cells in complex animal organisms: almost all active cell locomotion inside animal organisms is a migratory crawling movement (one exception is swimming sperm locomotion) [4]. During embryogenesis this migration of cells is essential for the structuring of the new organism – cells migrate to their eventual specific position in the developing organism, where they differentiate to their special purpose (like skin cells or neurons). Furthermore, cell migration is not only relevant for the development of multicellular organisms but also for maintaining this organism, ranging from wound healing to immune responses: Macrophages and neutrophiles crawl to infection sites and kill or disable foreign pathogenic agents as the central part of immune response. The remodelling and renewal of bones is mediated by osteoblasts, which crawl and tunnel into bone structure that is filled by other following osteoblasts. Similarly, fibroblasts migrate through connective organic tissues, allowing for remodelling the tissue if necessary or repairing damaged tissue [4]. Ultimately, cell migration (as an orchestration of the movement of different specialised cells to specific locations inside the organism) is an integral part for the organisation of animal organisms.

In vitro the migration of single mammalian tissue cells was first observed as early as 1675 when van Leeuwenhoek (24th October 1632 – 26th August 1723) saw cells crawl across his microscope slide during his microscopy studies. There are certain activities observable while the cell is migrating [2]:

- Protrusion – the cell is polarised, having a distinct shape with a front and a rear end. During migration the front end (also called the leading edge in scientific literature) is protruding, either in a large and flat front of protruding plasma membrane, the *lamellipodium*, or in many small tubular extensions called *filopodia*.
- Motion – the rest of the cell body including the nucleus seems to flow into the

direction of the leading edge, while there is also some retrograde flow observable from the front end towards the body centre [3].

- Retraction – at the rear the cell seems to disrupt from the surface and the plasma membrane is retracted towards the cellular body, indicating that there is some attachment or adhesion involved in the process of cell migration.

Thus, after pure observation of the behaviour of migrating animal tissue cells like fibroblasts or keratinocytes, the questions remain: how do the cells migrate over a surface, how do they generate the forces of traction and what biophysical evidence of the involved cellular mechanism of cell migration is available?

1.2.1. Biophysics of cell migration

The molecular mechanisms behind cell migration have become the focus of scientific research in the last decades, whereas the advances in fluorescence microscopy, molecular biology and biochemistry help to examine the involved processes of the motility and the discovery and identification of the molecular factors behind these processes. However, the experimental techniques cannot adequately explain which process or factor is generating the required forces for migration nor the exact physical mechanism. In vitro measurements of the exact forces generated by the involved proteins [31, 95, 18, 76] and in vivo measurements of the associated forces [73, 52, 20] brought a substantial advance in this direction. Theoretical biology and computational simulation and modelling enhance the gained experimental data and help with the quantification in proposing a model of the integrated cell migration mechanism and the generated forces beneath [6]. Cells experience external forces when moving on a substrate (e. g. in vivo the extracellular matrix inside the organism, or in vitro a microscopic slide). This includes the viscous forces of the surrounding medium and the interaction forces with the substratum. Internal forces include the forces generated by the cytoskeleton, which is required for coordination of the entire process, so that the inherent cytoskeletal forces are essential for this kind of motility.

The cytoskeleton is a mesh-like polymer network, comprising three types of biopolymers: actin, microtubules and intermediate filaments. Their difference to each other is their rigidity, described by their persistence length – the distance over which a filament is significantly bent by thermal forces – which increases with increasing stiffness [69].

Actin filaments are semiflexible polymers with a persistence length around 17 μm [42]. They have a diameter around 7 nm and are built from dimer pairs of globular actin monomers. Actin filaments are polar, they have two distinct ends: the plus end and the minus end [4]. The growing of actin filaments is dependent on the critical concentration of monomeric actin. The minus end has a six times higher critical concentration than the plus end (~0.6 μM and ~0.1 μM). When the monomeric actin concentration around the end of an actin filament is higher than the critical concentration, the monomeric actin is bound to the end of the filament and this end is growing by polymerisation. Conversely, when the concentration is below the critical concentration, actin monomers tend to detach from the filament and it is shrinking by depolymerisation. By having different critical concentrations at both ends of the filament, the actin filament can grow (or shrink) asymmetrically, in general the plus end is a fast growing end and the minus one a slow growing end. The moment, when the plus end grows and the minus end shrinks but the length of the filament stays more or less constant, is called treadmilling. This way the actin filament can transfer momentum forward. This is critical on how single actin filaments are able to generate forces.

Mictrotubules are the stiffest filaments, their persistence length lies in the range of 100 to 5000 µm depending on the length of the filament [74]. They have an outer diameter around 25 nm. This high rigidity is due to their hollow tube structure that is built by 13 protofilaments. These protofilaments are formed by tubulin protein subunits. The filament's polymerisation dynamics are similar to those of actin. Microtubules are polar, are able to treadmill, and can generate forces through polymerisation [30].

Intermediate filaments are much more flexible compared to actin filaments and microtubules, with a persistence length from ~0.3 to ~1.0 µm and in diameter between the others from 8 – 12 nm. They can be divided in different classes such as keratin, vimetin, desmin, lamin, neurofilaments, etc. Different cell types have normally different intermediate filaments. Contrary to actin filaments and microtubules, intermediate filaments are much more static, they are not polar, do not treadmill and normally do not depolymerise after polymerisation [109].

These biopolymers establish by different interconnections (such as crosslinking, bundling, binding by (motor) proteins or by simple entanglements) the internal mesh-like structure: the cytoskeleton. The coordination and combination of these polymers create a unique dynamic composite material that helps the cell with its structural integrity, shape, organisation of internal structures and internal transportation system, and – as already mentioned – cell motility.

The process of cell migration

Cell migration is a highly complex process dependent on the cytoskeletal cortex beneath the plasma membrane [4], and especially the actin cytoskeleton plays a predominant role in cell movement, as it is considered as the engine that drives cell protrusions and also mediating the required adhesion or detachment and the translocation of the cell.

Once cell migration begins, an integrated mechanism is constantly restructuring the actin cytoskeleton, which leads to the observable stages of cell migration [Figure 1.3]: First, the protrusion of the membrane forward (by orienting and reorganizing the actin network at the front). Second, the adhesion of the cell to the substrate at the leading edge and the detachment of the cell body at the rear. Third and finally, the contractile forces, generated largely by the successive interaction of the acto-myosin network, pull the cell

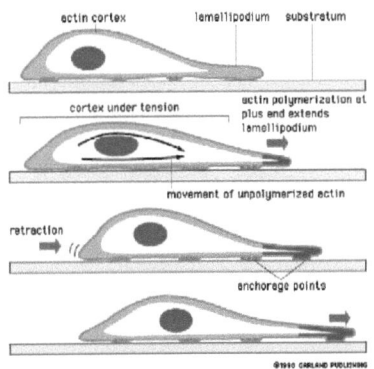

Figure 1.3: Schematics of a model of cell migration: *"The actin-polymerisation-dependent protrusion and firm attachment of a lamellipodium at the leading edge of the cell moves the edge forward (green arrows at front) and stretches the actin cortex. Contraction at the rear of the cell propels the body of the cell forward (green arrows at back) to relax some of the tension (traction). New focal contacts are made at the front, and old ones are disassembled at the back as the cell crawls forward. The same cycle can be repeated, moving the cell forward in a stepwise fashion. Alternatively, all steps can be tightly coordinated, moving the cell forward smoothly. The newly polymerised cortical actin is shown at the front."* from Alberts et. al.: *Molecular Biology of the Cell* [4].

forward [1].

In vivo actin filaments are assembled either as mesh-like networks or as bundles. Mesh-like actin filament networks are comprised by short crosslinked actin filaments. These actin filaments are primarily found at the leading edge during migration [52]. The growth of this mesh-like network of actin filaments, respectively the continuous re-creation of the actin meshwork, is mediated by the coordination of numerous accessory proteins [77, 78] as part of the signalling cascade: Activator proteins of polymerisation (such as the Arp2/3 complex, which is activated by the Wiskott–Aldrich Syndrome protein WASp, a mediator of the signal transduction [79]) as nucleator initialising the polymerisation and assembly of new actin filaments; cofilin (also known as Actin Depolymerisation Factor ADF) severs actin filaments and creates new plus ends; actin binding proteins (such as profilin and thymosine-β-4) provide a consistent pool of actin monomers; crosslinker and bundling proteins (such as filamin, α-actinin, fascin) connect the actin filaments to each other; capping proteins (such as CapZ) control the length of the filaments by attaching to the filament's end and stopping further polymerisation; severing proteins (such as gelsolin, serverin) cut actin filaments and their networks.

Actin bundles are composed of parallelly stacked and closely packed actin filaments, crosslinked by proteins such as fascin, fimbrin and scruin. They are known as stress fibres, responsible for force distribution across the cell and reinforcing adhesion sites [52]. They often connect distal points of adhesion so that tension can be propagated across the cell, especially the application of forces on the substrate for movement, fulfilling structural and sensory tasks for cell migration.

These different types of combined actin filaments are responsible for the different membrane structures observable at the leading edge of the cell. The protrusion of the cell membrane seems to be primarily based on the forces generated by actin polymerisation

1. Introduction

pushing the membrane outward [75]. Different types of cells have different types of protrusive membrane structures during migration, all are densely filled with actin filaments.

Filopodia (as seen by some types of fibroblasts) are essentially one-dimensional structures, consisting of a core of long bundled actin filaments, they function as chemical and mechanical sensors [111, 103]. Lamellipodia (formed by fibroblasts and other epithelial cells like keratinocytes and some neurons) are two-dimensional, sheet-like structures, containing an orthogonally cross-linked meshwork. A third type of protruding membrane structures, pseudopodia, is commonly formed by moving amoebae and neutrophils. These are three-dimensional temporal projections filled with a visco-elastic gel of actin filaments [4].

Lamellipodia are the best studied structure of cell protrusions during cell migration. They contain all the machinery required for migration. Studies with keratocytes taken from frogs and fishes (epidermal epithelial cells with rich abundance of keratin filaments) showed that these cells are moving very fast, up to 30 µm min^{-1}. In culture fragments of the lamellipodium can be sliced off with a micropipette. These fragments are able to continuously crawl on its own like a full cell, though lacking the organelles and microtubule network [4]. Marking a small patch of actin filaments in keratocyte lamellipodia reveals that the actin filaments inside remain stationary in relation to the substrate outside, while the lamellipodium crawls forward. The two-dimensional meshwork is formed by actin filaments which are mostly oriented with their plus end facing to the direction of movement, while the minus ends are attached to the sides of other actin filaments by ARP complexes. The whole meshwork seems to be treadmilling with growing free plus ends at the front and disassembling the meshwork at the minus ends at the back [4].

The nucleation and the growth of actin filaments is therefore located at the leading edge and directed to the plasma membrane, hence the growth and assembly of actin filaments at that location will push the plasma membrane forward. The main part of depolymerisation occurs well behind the leading edge, because cofilin preferably binds to actin filaments containing ADP-actin, whereas freshly assembled filaments contain ATP-actin. After ageing of the filament and ATP hydrolysis, cofilin disassembles most likely older filaments, whereas the fresh growing ends at the leading edge should be more or less resistant to depolymerisation by cofilin. This delayed ATP hydrolysis helps to establish a unidirectional process of treadmilling at the location of the lamellipodium [4].

But not only polymerisation forces play a role in motility, actin filaments are also able to generate motility forces through interaction with myosin, which can form bipolar myosin II filaments. Myosins are motor proteins and these myosin motors typically consists of a head, neck and tail region. The number of heads can be different, some myosin motors have one, other have two heads. This head/neck region is able to attach and to generate force, while the tail is most likely for anchoring to other filaments or cargo transportation. In connection with actin filaments (the so called acto-myosin complex) myosin motors are able to produce a three-step process of binding, power stroke and unbinding. The continuous repetition of this process generates a contractile force, which can push or pull actin filaments in a new orientation and is thought to be essential for pulling the whole cell body during migration [4].

This dynamic (re-)organisation of the cytoskeleton is one essential but not sufficient part of cell migration. For locomotion on a substratum, the cell must physically interact in some way with the surface it is moving on. The cell must be able to apply the contractile forces (that are generated by the cytoskeletal filaments) on the surface as a traction force. Hence, some degree of adhesion seems to be necessary for cell crawling, especially linking the cytoskeleton to the substratum by adhesion. But observation of cells shows that adhesion and locomotion seem to be inversely related: Highly adhesive cells move more slowly than weakly adhesive ones. Migrating cells have to find a balance between adhesion strength, forward actin protrusion and rearward actin flux [4].

Soon after the protrusion of the leading edge, the protruding plasma membrane (in form of a lamellipodium, filopodium or pseudopodium) forms new attachment sites with the surface. This is observable with interference reflection microscopy, where the cell's plasma membrane is extremely close to the substratum [4]. Transmembrane adhesion molecules seem to gather at the leading edge and the protrusion of the membrane is moving them forward to form new substrate attachments in front of the cell as new anchorage points. Once attached, these sites remain stationary and the cell moves over them until they appear at the rear of the cell. If adhesion fails, this adhesion complex is usually transported back on the dorsal surface of the cell as a membrane "ruffle" [4]. These transmembrane adhesion molecules (such as the integrin dimers found in fibroblasts and keratinocytes) are linked to the cytoskeletal filament meshwork and bundles inside the cell, allowing a direct connection of the interior network to the outer substrate. This linkage is created by certain adapter complexes such as α-actinin, talin, vinculin and others, which bind to the interior part of the integrin dimers and crosslinked actin filaments, while the outer part of the integrin dimer binds specifically to the surface. This way, the cytoskeletal contractile forces can be applied to the substratum (myosin II is mostly concentrated at the posterior of the cell [4]), and these traction forces exert a pull on the substratum. Besides, these attachments prevent the retraction of the protruding leading edge. In vivo, most moving cells in an animal organism crawl over a semiflexible protein mesh, the so called extracellular matrix, which is deformed and rearranged by these forces. Finally, the rest of the cell is pulled forward by these contractile forces, generated by the acto-myosin complex at the cell body and the rear. At the rear end the attachment must be destroyed, one possibility is by destroying the whole complex, another possibility by simple mechanical disruption of the complex or the loss of the adhesion site.

All the described processes are simultaneously and continuously running during cell migration. Based on cell migration experiments, this polarisation of cells and the direction of movement seems to be determined by locally occuring variations in the elasticity and rigidity of the actin network and hydrostatic pressure [16]. This is an important part during the entire process: the transition of the actin cytoskeleton between a solution-like viscous material – a sol – and a solid-like elastic material – a gel. It is most likely caused by the constant net polymerisation of actin and its network assembly at the leading edge and the simultaneous depolymerisation and disassembly at the rear end. These transitions lead to local changes in the cytoskeletal elasticity of the moving cell, generating some sort of visco-elastic gradient from front to rear (with a "sloppy" front end and a "stiff" rear end) that help to propel the migration. This increasing elasticity gradient can be explained by the stronger ordering and bundling of actin filament from front to rear. A single actin filament has a persistence length, describing its single rigidity. By bundling multiple filaments, their single rigidity is adding up, making a much stiffer bundle with a higher persistence length and rigidity. This essentially required gradient is confirmed by simulation studies and experimental data [6, 37].

Figure 1.3 summarises the entire process in a schematic view. Cell migration is coordinated temporally and spatially by many factors such as proteins and mechanical changes in the cytoskeleton and force generating structures. The other biopolymer filaments may have a supporting role in mediating cell migration: Microtubules radially extend from the centrosome to the actin network at the cell cortex in most cell types – forming some sort of a hub and spoke arrangement. These microtubules may aid in determining the directional cell movement [34]. Intermediate filaments form a network that spans the whole cell interior, providing cellular structural integrity. In general, they are more static in nature, making them less likely to be dynamically involved in cell migration. But some newer evidence suggest, that they might be more dynamic than previously thought [46].

The forces of cell migration

Protrusion mechanics Protrusion is the initial step of cell migration. Although the integrated machinery behind this step is very complex, the protrusion of the leading edge is simply believed to be caused by the polymerisation of actin. Polymerising actin filaments can generate a significant force (without motors), but to protrude a membrane, this polymerisation force has to be applied against a load – the plasma membrane. The question is on how a filament generate such a force. There are two main models trying to answer this question by explaining the generated polymerisation forces and the accompanied protrusion of membrane: the ratchet model [66, 67] and the autocatalytic model [23, 22]. The problem is, if the membrane is like an immovable wall, polymerising actin filaments would stop growing after bumping into this wall, unable to generate a force pushing against it.

The ratchet model considers the membrane fluctuating under Brownian motion – small thermal fluctuations because of its small size and flexibility. Additionally, an actin filament is also flexible and can bend in response of load, allowing actin monomers to insert itself in the small gap between the filament end and the membrane. The filament is able to grow and generate an elastic force, pushing the membrane away. It is working like a ratchet, implying the prevention of backward movement of the membrane with a small forward movement of the cell edge. This is the *Elastic Brownian Ratchet Model* as an ex-

1.2. Biological cell locomotion

planation of the force generated by single polymerising actin filaments [66]. An extension of this model is the *Tethered Elastic Brownian Ratchet Model*. It distinguishes between "working" filaments (which are able to apply a force) and "non-working" filaments (which are attached to the surface and not able to apply a force) [67]. Both are able to switch to the other type. Force is generated by multiple filaments in a branched actin network, whereas new filaments and branches are polymerised independent of existing branches, following the Dendritic-Nucleation Model.

Theoretical calculations of this ratchet mechanism indicate that the maximum force generated by a single actin filament is about 5 – 7 pN [68]. Considering this with the assumption of hundreds of actin filaments pushing the leading edge will result in a force of nanonewtons per micrometer, enough to cope with the membrane load and resistance [67]. These force calculations do not include contractile forces of motor proteins, therefore even higher forces are possible by incorporating motor proteins which are able to convert free energy to work in form of contractile forces.

The autocatalytic model also tries to explain the forces generated by actin polymerisation (with a similar basis as the ratchet models) with two approaches: a numerical approach [23], and a deterministic approach [22]. Both approaches assume that new actin branches are generated from existing branches – in contrast to the Tethered Elastic Brownian Ratchet Model. The attachment of filaments to surface is also not considered by these approaches.

Both the ratchet and the autocatalytic models predict an exponential relation between force (or load) and actin growth velocity. However, their underlying model assumptions and differences in actin branching assumptions and handling of actin filament orientation make both fundamentally different models [100]. The predicted force-velocity relations have been tested experimentally in vitro [61, 107]. Further in vitro motility experiments and the gained data were not able to favour one model but revealed more insights of the underlying forces and principles [7].

In vivo experiments are even more difficult to test force-velocity relationships for ratchet or autocatalytic models as there are more factors which need to be considered. As example the hydrostatic pressure of the cell might play a role [6]. Another example is that membrane resistance is influencing the protrusion velocity [65]. Results of further experiments suggest an inverse relation between membrane resistance and protrusion [80].

Recapitulating, neither the ratchet nor the autocatalytic models has been proven or ruled out as a model for the molecular mechanisms of force generation of actin filaments, both suggest a likely mechanism and help to understand the actual protrusion mechanism. There is even the possibility of a combination of both models working in vivo, hence further studies might give a clue about all the parameters required to understand actin-driven membrane protrusion.

1. Introduction

Adhesion mechanics The adhesion mechanics involved with cell migration and the applied traction force are also under investigation. Different cell types have a different spatial distribution of attachment sites and differ in adhesion forces. They are important factors of the rate of protrusion and translocation speed [56, 70]. Near the leading edge adhesions are required as anchorage points to convert polymerisation forces into protrusion. Simultaneously, polymerising actin is also flowing away from the leading edge, known as centripetal actin flow [49, 62].

There is one phenomenon observable in many migrating cells (such as fibroblasts, fish keratocytes or neuronal growth cones): the *retrograde flow*. It describes the movement of actin filaments rearwards in the cell in opposite direction to movement [59, 58, 51]. Experiments have shown that actin polymerisation is one crucial factor of the generation of retrograde flow by providing a constant source of actin and pushing (or pulling) the actin network of the lamellipodium backward [38, 106], whereas myosin motors are important for retrograde flow as shown in fish keratocytes [101], in growth cone movement [29, 19, 105] and fibroblasts [21]. In summary, these studies indicate that both myosin motors and actin polymersation are crucial for driving the retrograde flow [47], but to which extend the one is more critically important than the other may be different in different cell types.

The relation between translocation speed and retrograde flow is linked by a molecular clutch [51, 64]. This molecular clutch consists of vinculin, talin and other adhesion complexes. It determines the strength of the bond and linkage of the inner cytoskeleton to the outer substratum. This interaction is crucial to transmit the contractile forces generated by the cytoskeleton onto the underlying substratum. It enhances the translocation speed and decreases the retrograde flow, because the contractile forces caused by asymmetry are applied now on the substratum causing traction forces and allowing the cell to pull itself forward by pushing against the substrate, because the sum of the horizontal traction force components has to vanish. Without this clutch there is only a loose connection between the cytoskeleton and the substratum and force is applied ineffectively accompanied with a high retrograde flow. The actin network is pushing itself backward because of the lesser resistance instead of applying its contractile forces on the substrate. This results in a lower translocation speed.

Therefore the "clutch hypothesis" tries to predict that slow moving cells have a high retrograde flow but a low translocation speed due to lesser traction forces, whereas it is vice-versa in fast moving cells, low retrograde flow but large traction force, causing a high translocation speed. Though, experimental studies were not able to confirm this hypothesis. There seem to be more complex relations between translocation speed, traction forces and retrograde flow, which are revealed by further studies. A phenomenon called adhesion raking – describing the raking inward of the cytoskeleton against the substrate – is also able to produce a retrograde flow [51]. This adhesion raking and the clutch disengagement may occur simultaneously and their combined effects might produce a non-linear biphasic relation between translocation speed and traction forces [51]. The

1.2. Biological cell locomotion

translocation and traction relation is therefore non-linear – this seems to be confirmed by other experimental and computational studies [6, 94, 88, 110, 44, 55]. The conclusion: in fast moving cells the adhesion forces are at their optimum and the retrograde flow is minimised, while in slow moving cells the adhesion forces are below or above optimum and the retrograde flow is high.

Experiments to obtain an estimation of the adhesion force – the force required to break a single integrin adhesion – revealed a force value of ~10 – 30 pN [96, 41]. As mentioned before, the polymerisation force of an actin filament is ~5 – 7 pN. Estimating a dozen actin-integrin bonds per μm, the total force generated by the leading edge is likely to be several nanonewtons. Studies measured around 34 – 85 nN [20]. The magnitudes of various measured forces in different cells are also available [7].

The adhesion of the leading edge is normally attended by the de-adhesion of the cell body at the rear, most likely by a biochemical disassembly of the focal adhesions, mediated by numerous proteins including the protease calpain and signal proteins such as Src, FAK and PAK, etc. [81, 82, 83]. Another process of detachment might be simple mechanics – disruption by the contractile force. The breakage is essential for movement, otherwise the cell is not able to move further because the bonds are working as anchor [6].

Recapitulating, adhesion forces are a crucial part of cell migration, essential for the application of traction forces on the substratum, which is necessary for translocation.

Retraction mechanics The final step of cell migration: the retraction of the rear of the cell body. The retraction force involved for this process is most likely generated by the acto-myosin complex by sliding myosin motors. Several studies have explored the role of myosin in retraction [54, 50, 93, 102], an imbalance of adhesion and contraction is lifting the rear end, concluding that myosin is an important factor of the retraction mechanics. Studies on the acto-myosin complex and its contractile force revealed that it has a similar magnitude of the polymerisation force of the leading edge. Each myosin motor is generating ~1 pN of force, with thousands of myosin motors at the rear [68].

The acto-myosin seems to be the primary factor for the generation of retraction forces, but other processes might also be involved. One possible factor is the solation of the actin network at the rear – the transition from a solid-like elastic material – a gel – to a solution-like viscous material – a sol. Some evidence suggests that this "fluidisation" – the transition to a more fluid-like state – may also be driven by molecular motors [48]. The transition can be explained the following: At the rear are filament bundles. Each filament has a persistence length and rigidity. The filament bundle has also bundled the rigidity of each individual filament, making the bundle much stiffer than the single filament. When a filament detaches from the bundle, its persistence length and rigidity decreases, it contracts because of the increase in entropy. The rigidity is further decreased by disassociation of the monomers during depolymerisation.

Recapitulating, the retraction force is most likely caused by cytoskeletal disassembly and contractile forces generated by the acto-myosin complex. It is a required force to close the process cycle of cell migration.

1.2.2. Bionic abstraction

Cell migration is a complex integrated cellular process involving many molecular factors. It seems rather difficult or nearly impossible to technically rebuild this mechanism for usage as a technical locomotion device. Fortunately, bionics is not about copying or rebuilding nature. Bionics is using a virtual creative tool for creating technological applications inspired by nature. This virtual tool is the ability of the human mind for abstract thinking. The process of abstraction helps to see the essential principle behind, which then can be transferred into a technical application. Simplification is one type of abstract thinking.

Hence, for a robotic locomotion device inspired by cell migration there is no need to rebuild the exact molecular mechanisms in a technological manner. The cell and its migration process just need to be simplified. Lets start with the simplification of a cell. The simplest model of a cell is a water-filled balloon. The elastic skin of the balloon represents the sub-membranous cell cortex (the membrane with the cytoskeletal layer beneath), which has certain elastic properties. The water in the balloon represents the cytosol. In relaxation the water-filled balloon has the shape defined by the force equilibrium of the inner forces (such as water pressure) and outer forces (such as air pressure), but like a cell this water-filled balloon is deformable, forces acting on the balloon are able to change its shape by squeezing or stretching. Unfortunately, a water-filled balloon is not able to actively move on its own. It may use its potential energy by rolling down a slope or falling from height, but nothing more. So the balloon requires additional "props", something the cell has and the balloon has not. Of course, a cell has structure, it is not just filled with water. Water is just the solvent and reaction room for all the organelles and molecular structures inside. This structure is the cytoskeletal cortex with its filamentous actin. This cortex has visco-elastic properties and these properties can be altered by dynamically reorganising this structure – the basic principle of cell migration.

For the bionic locomotion device, the most simple required model of a cell is a cross-section of an elastic deformable membrane sack, filled with a liquid and structured with a cytoskeletal cortex [Figure 1.4a] with some mechanical properties. This continuous model is more simplified by discretisation. The shape of the cortex and the membrane can be described as a polygon with vertices that are connected by segments, and the mechanical properties of the cortex can be implemented into the segments and vertices, which are usable as joints for the segments. The result of this simplification is a very simple discretised mathematical model of a cell [Figure 1.4b]. Since this model does look more technological than biological, this cell model can now be easily used as a simple robot model.

Only one thing is missing to let this model actively move: the motorisation. The motor of cell migration is driven by the simultaneous cytoskeletal reorganisation. Of course,

1.2. Biological cell locomotion

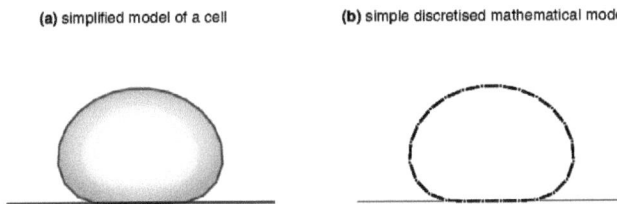

Figure 1.4.: A simplified cross-section model of a cell with a membrane in black and the elastic cytoskeletal cortex as ring beneath the membrane. This model can be converted in a simple discretised mathematical model, a chain composed of elastic segments connected via joints. The mechanical properties of the cytoskeletal cortex can be modelled as mechanics of the chain segments and the corresponding vertices.

polymerisation and depolymerisation of molecule-sized actin filaments is hardly realisable on a larger technology scale. The simplifying process of abstract thinking is also helping here. During cell migration actin is constantly polymerised into filaments that are further branched and bundled. This is causing a local transition of the cytoskeleton from a sol, a solution-like viscous material, to a gel, a solid-like elastic material. The cell is just generating a visco-elastic gradient from the front to the rear during cell migration. It has a "sloppy" end at the front and a "stiff" end at the rear. In a simplified view this gradient is the basic motor of cell migration. Additionally, this process is initialised and further enhanced by adhesion to the surface. Hence, the only requirement as motorisation for the mathematical cell/robot model is an adhesion induced mechanism, which is temporally and gradually changing the elastic properties of the attached parts of the chain, which simply return back to the previous elastic properties after detachment.

The questions on how such a mechanism can be modelled in detail and how this simple cell model is a basis for more advanced robot models utilising this movement principle and what capabilities do such types of robots have, are elaborately answered in the next chapters.

2. Modelling

THIS work introduces three different computational robot models with implementation of the biophysical locomotion principle of migratory cells, based on the biological background and the biophysical properties described in the introductory chapter. The model "cell" robots need to fulfil the requirement to adhere on a surface and the ability to build up an elasticity gradient of over time during adhesion. Furthermore, they are simplified as much as possible to emphasise their constructability and to enhance the bionic abstraction process. This chapter describes the different models and in-detail the involved forces, mechanics and dynamics. The shape of each model is described by its two-dimensional cross section -- simulation is done with the aid of the numerical computing environment and fourth-generation programming language MATLAB®. Part of this modelling is build upon the diploma thesis from November 2008 [11].

Additionally, this chapter provides insights into modelling randomly generated surface structures that are used for probing the robot models' behaviour on rough surfaces with different properties. The last section of this chapter recapitulates the intrinsic parameters used in modelling.

2.1. Introduction of robot models

Each robot model consists of flexible vertices connected by straight elastic segments. The core element of each model is at least one closed chain of such vertices and segments. Forces are finally calculated only for each vertex, resulting in a differential equation for the displacement of each vertex. "Outer" vertices are able to adhere to a given surface. Adhesion is a stimulus for dynamic adaptation by triggering a temporal change of chain or segment stiffness properties, whose details are different for each model. The change in stiffness represents the motor of the models, because energy is required for this stiffness adaptation, which is provided unlimited during simulation, so that every model is able to sustain a stable movement. Disruption of the bound vertices is caused by exceeding a certain force limit, thereafter the changed stiffness properties passively revert back to the free and unbound state [for more details about this driving mechanism, see subsection 2.2.3].

2. Modelling

(a) single-chain model

(b) extended single-chain model

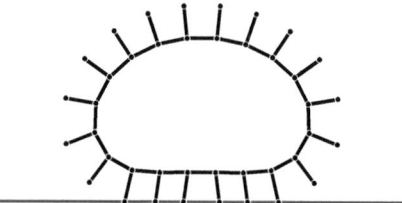

(c) double-chain model

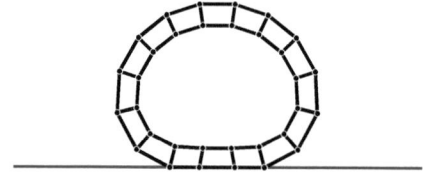

Figure 2.1.: The three robot models.

2.1. Introduction of robot models

(a) single chain model

The single chain model is the simplest of the three models and consists of one closed chain with 24 vertices connected by elastic segments [Figure 2.1a]. It is the most simplified abstraction of a cell with just a sub-membraneous cortex with elastic properties [cf. Figure 1.4]. Free and unbound vertices and the corresponding segments have a positive equilibrium curvature, setting a convex equilibrium shape of the chain. Adhesion triggers an alteration of the chain behaviour: bound vertices and the corresponding segments temporally and gradually decrease to a negative curvature, resulting in a local tendency to form a concave shape at the sites of adhesion. After disruption this bending tendency reverts slowly back to the initial free positive value.

(b) extended single chain model

The extended single chain model consists of a closed chain with 20 vertices connected by elastic segments and having extensions with elastic spikes at each vertex [Figure 2.1b]. It is a simplified abstraction of a cell, where the spikes may represent filopodia or pseudopodia. These spikes tend to align to the outer angle bisector of neighbouring chain segments. The outer vertices at the end of the spikes are able to adhere to the surface. Adhesion triggers the temporal change of the local chain equilibrium curvature alignment (as in the single chain model) and, additionally, the spikes also change their elastic properties: during adhesion their free length decreases over time, so that the spike becomes stiffer the longer it is bound to the surface. Moreover, the alignment to the angle bisector is loosened during adhesion. All temporally changed mechanical properties slowly revert back to their non-adhesion states after disruption of the spike's end.

(c) double chain model

The double chain model consists of two chains, an inner chain and an outer chain with 16 vertices each [Figure 2.1c]. The corresponding vertices of both chains are connected by radial elastic spoke segments that tend to align to the angle bisector. It is a simplified abstraction of a cell, where the compartments of the chains represent the elastic actin cortex, while the spokes correspond to the binding of the actin cortex to transmembrane receptor proteins. The inner chain segments are modelled to be much stiffer than the outer chain segments. Additionally, the inner chain has a bending stiffness and tends to straighten. Adhesion in this model triggers the temporal change of the elastic properties of the radial segments: during adhesion their free length decreases and the alignment of the radial segments to the angle bisector is loosened during adhesion. The temporally changed mechanical properties slowly revert back to their non-adhesion state after disruption. The double chain model is the main part of the diploma thesis from November 2008 [11].

2. Modelling

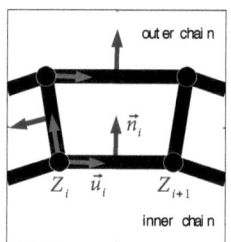

Figure 2.2: Configuration of the vertices and segments and each segment's unit vectors $\vec{u}$ and $\vec{n}$ in tangential and normal direction respectively, shown on the upper section of the starting configuration of the double chain model (respective parts are missing in the other two models, e. g. the outer chain segments and vectors).

2.1.1. Configuration and notation

The chain vertices Z_i ($i = 1, ..., n$) in each model are arranged clockwise and are defined by x- and y-coordinates. Each elastic segment between two vertices has a length l_i characterised by a tangential unit vector $\vec{u}_i$ (pointing from Z_i to Z_{i+1}) and a corresponding normal unit vector $\vec{n}_i$ (orthogonal counter-clockwise to $\vec{u}_i$).

In case of radial segments as in the extended single chain model and the double chain model, the tangential vector of the radial segment is pointing from the vertex of the inner chain to the outer vertex and the normal vector is orthogonal counter-clockwise to the tangential vector.

The outer chain of the double chain model is configured as described above. Figure 2.2 is showing the full configuration of the double chain model. In case of the other two models, respective parts are missing.

2.1.2. Non-dimensionalisation

For constructability purposes, the parameters of the models are non-dimensionalised. Example: the length l is divided by the free length l_0 giving the non-dimensional length $L = \frac{l}{l_0}$. The same goes for the coordinates of each vertex, dividing by l_0 gives non-dimensional coordinates. Non-dimensional parameters allow to scale the model during construction later, which is then only dependent on the desired size and the available materials.

2.2. Model mechanics

This section introduces in detail all force components, different friction models, and the adaptive dynamics of each of the three models – how an elasticity gradient is implemented that is responsible of the models' propulsion.

The vertex displacement $d\vec{Z}$ of each vertex in the model during a simulated time step dt is defined by a differential equation (giving the instantaneous vertex velocity $\vec{v}$):

$$\frac{d\vec{Z}}{dt} = \mathbf{H}^{-1} \cdot \vec{F}_Z = \vec{v} \qquad (2.1)$$

as the force resultant $\vec{F}_Z$ at the vertices transformed by the friction matrix $\mathbf{H}$ defining the part of the friction force that affects each vertex [subsection 2.2.2], while friction is only applied to the free moveable vertices. Additionally, normal distributed white noise can be stochastically applied to the displacement of the vertices. The models are assumed as an overdamped system, means that acceleration can be neglected in comparison to the friction involved: $m \cdot |\ddot{z}| \ll |F|$.

2.2.1. Static forces

The force resultant $\vec{F}_Z$ at each vertex is defined by the sum of the following forces:

Gravity

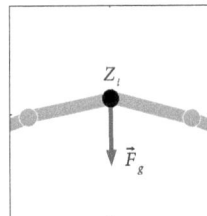

The gravity force F_g is given by a vector $\vec{g}$ representing the direction and magnitude a vertex is accelerated by gravity and a scaling factor f_g incorporating the non-dimensional "mass" of a vertex:

$$\vec{F}_g = f_g \cdot \vec{g} \qquad (2.2)$$

(see figure on the left).

Elasticity

Each segment between two nodes is elastic and modelled as a spring. The elastic force F_e of a segment is characterised by Hooke's law: the difference ΔL of the spring's length L to its free length L_0 and the spring constant k_e:

$$F_e = -k_e \cdot \Delta L \qquad (2.3)$$

2. Modelling

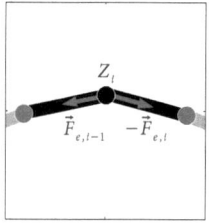

In the simple case of a vertex with two neighbouring segments, the total elastic force $\vec{F}_{e,Z_i}$ at this vertex is the sum of the elastic forces of both segments:

$$\vec{F}_{e,Z_i} = \vec{u}_{i-1} \cdot F_{e,i-1} - \vec{u}_i \cdot F_{e,i} \qquad (2.4)$$

and the tangential unit vector $\vec{u}$ defines the direction of the elastic force (see figure on the left).

Pressure

In each model a pressure force is applied, that counteracts the deformation of the model's shape, stabilising the general shape and preventing a collapse. The pressure force $\vec{F}_p$ is defined by ΔA, the difference of the area A to the equilibrium area A_0 (in this case the area is equivalent to the volume, because of the two-dimensionality of the model, thus the change of area is equivalent to a change of volume which is causing a pressure) acting on each segment's length L multiplied with k_p, defining the "constant" pressure per volume change:

$$F_p = -k_p \cdot L \cdot \Delta A \qquad (2.5)$$

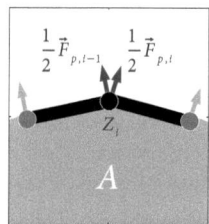

Pressure is a force exerted on an area (in this case the area is simply the segment's length, because of the two-dimensionality of the models). The portion of the pressure force (which is acting on the vertices on the ends of the segment) is therefore the half of the force. In the simple case of a vertex with two neighbouring segments, the total pressure force $\vec{F}_{p,Z_i}$ at this vertex is the sum of the halves of both neighbouring segments' pressure forces:

$$\vec{F}_{p,Z_i} = \vec{n}_{i-1} \cdot \frac{1}{2} F_{p,i-1} + \vec{n}_i \cdot \frac{1}{2} F_{p,i} \qquad (2.6)$$

and the normal unit vector $\vec{n}$ defines the direction of the pressure force (see figure on the left).

In case of the double-chain-model, each peripheral compartment enclosed by the segments of the inner and outer chain and the radial segments has an additional pressure force applied on the four neighbouring segments of that chamber, proportional to its volume change.

Moment of force

The inner chain is modelled with a bending stiffness, realised by a torsion spring at each vertex trying to set the neighbouring segments on a predefined angle. This mechanism is

2.2. Model mechanics

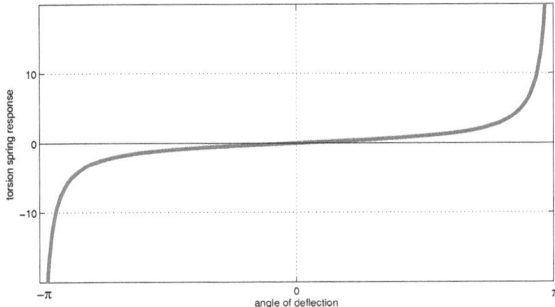

Figure 2.3.: Torsion spring response to the deflection angle of the spring. The deflection angle has to be limited between $-\pi$ and π, because the torsion spring response is getting infinite close to these limits, preventing an overexpansion of the spring.

causing a moment of force on each segment. The segment's moment M is characterised by the deflection $\Delta\varphi$ of the angle φ to its equilibrium angle value φ_0 and a spring constant k_m. The spring's response to the angle deflection is not linear but given by $\tan(\frac{\Delta\varphi}{2})$ (to prevent the overexpansion of the torsion spring [Figure 2.3]):

$$M = -k_m \cdot \tan(\frac{\Delta\varphi}{2}) \qquad (2.7)$$

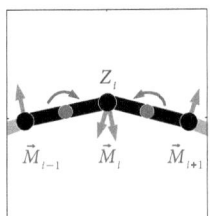

In the simple case of a vertex with two neighbouring segments, the torsional force $\vec{F}_{m,Z_i}$ at this vertex is the sum of the moments, considering that a moment of force is acting as a lever on both ends of each segment over the segment's length L (therefore each segment is basically a lever):

$$\vec{F}_{m,Z_i} = \vec{n}_{i-1} \cdot \frac{1}{L_{i-1}}(M_i - M_{i-1}) + \vec{n}_i \cdot \frac{1}{L_i}(M_i - M_{i+1}) \qquad (2.8)$$

and the normal unit vector $\vec{n}$ defines the direction of the moment (see figure on the left).

In case of the extended single-chain model and the double-chain-model an additional moment of force is acting on each radial segment that aligns these to the angle bisector [see subsection 2.2.3 for more details].

2.2.2. Friction forces

Friction is a force resisting the relative motion of elements against other elements, such as the surrounding viscous medium. Depending on the environment and the involved elements friction can be categorised in several types, e. g. dry kinetic or static friction

2. Modelling

between two moving or non-moving surfaces, viscosity of fluids or drag between solid objects and a fluid (or gas).

Simple friction model

In this work the simplest model of friction implies that the friction force $\vec{F}_{f,i}$ of a vertex is proportional to its velocity $\vec{v}_i$ with friction coefficient η defining the magnitude of friction. For a single chain, this is given by the formula:

$$\vec{F}_{f,i} = -\eta \cdot \vec{v}_i \tag{2.9}$$

though this friction model is quite simple but generally not realistic, only in cases where the vertices have relatively large volumes.

In this case the friction matrix **H** [Equation 2.1] is a $n \times n$ scalar matrix with η in the main diagonal (with n the number of vertices of the chain):

$$\mathbf{H} = \begin{bmatrix} \eta & 0 & \cdots & 0 \\ 0 & \eta & & \vdots \\ \vdots & & \ddots & 0 \\ 0 & \cdots & 0 & \eta \end{bmatrix}$$

Drag dependent friction model

For improvement towards a more realistic friction model the assumption is made, that any moving segment is causing drag, which counteracts the velocities of the two vertices of this segment. For the single chain, this drag force is given by the formula:

$$\begin{aligned}\vec{F}_{f,i} = &- [\frac{\eta_n}{12}[L_{i-1}(\vec{n}_{i-1} \cdot \vec{n}_{i-1}^T)] + (\frac{\eta_u}{2} + \eta_e)[(\vec{u}_{i-1} \cdot \vec{u}_{i-1}^T)]] \, \vec{v}_{i-1} \\ &- [\frac{\eta_n}{6}[L_{i-1}(\vec{n}_{i-1} \cdot \vec{n}_{i-1}^T) + L_i(\vec{n}_i \cdot \vec{n}_i^T)] + (\frac{\eta_u}{2} - \eta_e)[(\vec{u}_{i-1} \cdot \vec{u}_{i-1}^T) + (\vec{u}_i \cdot \vec{u}_i^T)]] \, \vec{v}_i \\ &- [\frac{\eta_n}{12}[L_i(\vec{n}_i \cdot \vec{n}_i^T)] + (\frac{\eta_u}{2} + \eta_e)[(\vec{u}_i \cdot \vec{u}_i^T)]] \, \vec{v}_{i+1}\end{aligned} \tag{2.10}$$

involving the velocity of the vertex and the two neighbouring vertices (two neighbouring vertices define the motion of one segment). The drag of the segment is projected in tangential and normal direction of the segment with corresponding coefficients η_u for tangential and η_n for normal direction. Vectors marked with T are transposed vectors. Additionally, the internal friction by elastic deformation of the segment is defined by η_e [for full derivation of the formula, see section A.1].

In this case the friction matrix **H** [Equation 2.1] is a cyclic tridiagonal $n \times n$ block matrix (with 2×2 matrix blocks, because of the multiplication of vector with transposed vector) with the 2×2 matrices $\mathbf{D}_i$ for $\vec{v}_i$ in the main diagonal and the 2×2 matrices $\mathbf{C}_i$ for $\vec{v}_{i-1}$

and E_i for $\vec{v}_{i+1}$ in the secondary diagonals (with n the number of vertices of the chain):

$$H = \begin{bmatrix} D_1 & E_1 & 0 & \cdots & 0 & C_1 \\ C_2 & D_2 & E_2 & & & 0 \\ 0 & C_3 & D_3 & E_3 & & \vdots \\ \vdots & & \ddots & \ddots & \ddots & 0 \\ 0 & & & C_{n-1} & D_{n-1} & E_{n-1} \\ E_n & 0 & \cdots & 0 & C_n & D_n \end{bmatrix}$$

Moment dependent friction model

The third friction model is assuming that friction resistance is due to motion within in the torsion springs, that are actively aligning the chain segments. Therefore the friction is dependent on the temporal change of the angle between the two segments and defined as a moment of force at this vertex, embodied in the following formula:

$$\begin{aligned}
\vec{F}_{f,i} = & -[\tilde{\eta}_{i-1} \frac{1}{L_{i-1}} \vec{n}_{i-1} \cdot \vec{m}^+_{i-1}] \vec{v}_{i-2} \\
& + [\tilde{\eta}_i (\frac{1}{L_i} \vec{n}_i + \frac{1}{L_{i-1}} \vec{n}_{i-1}) \cdot \vec{m}^+_i + \tilde{\eta}_{i-1} \frac{1}{L_{i-1}} \vec{n}_{i-1} \cdot (\vec{m}^+_{i-1} + \vec{m}^-_{i-1})] \vec{v}_{i-1} \\
& - [\tilde{\eta}_i (\frac{1}{L_i} \vec{n}_i + \frac{1}{L_{i-1}} \vec{n}_{i-1}) \cdot (\vec{m}^+_i + \vec{m}^-_i) + \tilde{\eta}_{i+1} \frac{1}{L_i} \vec{n}_i \cdot \vec{m}^+_{i+1} + \tilde{\eta}_{i-1} \frac{1}{L_{i-1}} \vec{n}_{i-1} \cdot \vec{m}^-_{i-1}] \vec{v}_i \\
& + [\tilde{\eta}_i (\frac{1}{L_i} \vec{n}_i + \frac{1}{L_{i-1}} \vec{n}_{i-1}) \cdot \vec{m}^-_i + \tilde{\eta}_{i+1} \frac{1}{L_i} \vec{n}_i \cdot (\vec{m}^+_{i+1} + \vec{m}^-_{i+1})] \vec{v}_{i+1} \\
& - [\tilde{\eta}_{i+1} \frac{1}{L_i} \vec{n}_i \cdot \vec{m}^-_{i+1}] \vec{v}_{i+2}
\end{aligned}$$

(2.11)

with $\vec{m}^+_i = \frac{1}{L_{i-1}}[\vec{n}_i + \vec{u}_{i-1}(\vec{n}_{i-1} \cdot \vec{u}_i)]$ and $\vec{m}^-_i = \frac{1}{L_i}[\vec{n}_{i-1} - \vec{u}_i(\vec{n}_{i-1} \cdot \vec{u}_i)]$ and $\tilde{\eta} = \eta \cdot \frac{1}{\cos(\varphi)}$, involving the vertex velocity and the velocities of the next two vertex neighbours on each segment, because three moments are acting at one vertex [Equation 2.8] involving five vertices [for full derivation, see section A.1].

In this case the friction matrix H [Equation 2.1] is a $n \times n$ cyclic pentadiagonal matrix with the prefactor d_i for $\vec{v}_i$ in the main diagonal and the prefactor c_i for $\vec{v}_{i-1}$ and e_i for $\vec{v}_{i+1}$ in the secondary diagonals and prefactors b_i for $\vec{v}_{i-2}$ and f_i for $\vec{v}_{i+2}$ in the tertiary diagonals

2. Modelling

(with n the number of vertices of the chain):

$$H = \begin{bmatrix} d_1 & e_1 & f_1 & 0 & \cdots & 0 & b_1 & c_1 \\ c_2 & d_2 & e_2 & f_2 & & & 0 & b_2 \\ b_3 & c_3 & d_3 & e_3 & f_3 & & & 0 \\ 0 & b_4 & c_4 & d_4 & e_4 & f_4 & & \vdots \\ \vdots & & \ddots & \ddots & \ddots & \ddots & \ddots & 0 \\ 0 & & & b_{n-2} & c_{n-2} & d_{n-2} & e_{n-2} & f_{n-2} \\ f_{n-1} & 0 & & & b_{n-1} & c_{n-1} & d_{n-1} & e_{n-1} \\ e_n & f_n & 0 & \cdots & 0 & b_n & c_n & d_n \end{bmatrix}$$

A phenomenon of this moment dependent friction: it is possible that the moments of force are acting with the same magnitude in opposite directions with the result that the friction forces may cancel each other and no net friction could be applied at some vertices, causing instability during simulation. Indeed, H is not invertible, because $\det(H) = 0$, therefore this kind of friction model is used only in addition as enhancement for the previously presented drag dependent friction model.

2.2.3. Dynamics

The displacement $d\vec{Z}$ and the velocity $\vec{v}$ of a vertex is given by a differential equation [Equation 2.1] involving the force equilibrium at each vertex. The dynamic adaptation (the build-up of the elastic gradient) of each models is changing this equilibrium, resulting in a stable motion of the whole system. This adaptation is triggered by attachment: In the model, if a vertex is touching the ground surface, it will automatically attach and adhere to this surface and the displacement of this vertex is set to zero. To detach this vertex a force must pull it away from the surface:

$$\vec{F}_Z^\perp \geq f_h \qquad (2.12)$$

where $\vec{F}_Z^\perp$ is the vertical force component in relation to the surface and f_h the force height threshold. Therefore the vertical force component must exceed a certain predefined limit, which then detaches the vertex. Additionally, during adhesion the elastic properties of the neighbouring segments are changed temporally with a predefined rate. This is causing the required elasticity gradient: the motorisation of the model, because energy is required for actively triggering and changing the elastic properties. After disruption it reverts back passively to the previous state with a slower rate. The dynamic adaptation is introduced in detail for each model:

Single chain model

The elastic gradient in the single chain model is caused by a temporal change of the elastic properties of the chain's torsion springs, in other words: the predefined equilibrium angle between two neighbouring segments is changed. At a free vertex the torsion spring is set "short", so the segments tend to bend inwards (and the free vertex is pushed outwards). During adhesion this behaviour changes temporally and gradually in the opposite direction: the torsion spring is set "long", the segments tend to bend outwards (and it tries to pull the adherent vertex inwards). Thus, during adhesion the equilibrium angle φ_0 changes to a new value φ_{0max} over time with rate $r_{\varphi,a}$:

$$\frac{d\varphi_0}{dt} = (\varphi_{0max} - \varphi_0) \cdot r_{\varphi,a} \qquad (2.13)$$

After disruption the equilibrium angle φ_0 reverts back to value $\varphi_{0min} < \varphi_{0max}$ over time with rate $r_{\varphi,f}$:

$$\frac{d\varphi_0}{dt} = (\varphi_{0min} - \varphi_0) \cdot r_{\varphi,f} \qquad (2.14)$$

The change of the equilibrium angle φ_0 is influencing the moment of force M at this vertex [Equation 2.7] and changing the vertex's force equilibrium.

Extended single chain model

The elasticity gradient in the extended single chain model is caused by a temporal change of three elastic properties.

First, during adhesion of the outer vertex the free length L_0 of the radial segment decreases to a minimum free length L_{0min} (becoming stiffer over time) with rate $r_{L,a}$:

$$\frac{dL_0}{dt} = (L_{0min} - L_0) \cdot r_{L,a} \qquad (2.15)$$

After the vertex's disruption the free length of the radial segment reverts back to a maximum free length $L_{0max} > L_{0min}$ with rate $r_{L,f}$:

$$\frac{dL_0}{dt} = (L_{0max} - L_0) \cdot r_{L,f} \qquad (2.16)$$

Second, the torsion spring and the predefined equilibrium angle φ_0 change temporally and gradually as in the single chain model [see Equation 2.13 and Equation 2.14 in the previous section]. It is linked to the temporal change of the free length of the radial segments and uses the same rates.

Third, the alignment of the radial segments to the angle bisector is loosened during adhesion. It is linked to the temporal change of the free length of the radial segments and uses the same rates: in this case the spring constant k_m of the torsion spring (responsible for the alignment to the angle bisector) is softened during adhesion, decreasing to one fifth of the strength and reverting back to full strength after disruption [Equation 2.7].

The change of the free length L_0 of the radial segments is influencing the elastic forces F_e at the two neighbouring vertices [Equation 2.3] and the change of the equilibrium angle φ_0 is influencing the moment of force M at this vertex [Equation 2.7], both influencing the force equilibrium of the vertices.

2. Modelling

Double chain model

The elastic gradient in the double chain model is caused by a temporal change of the elastic properties of the radial segments. It uses the same mechanics as the extended single chain model, the free length of the radial segments are adapting to the state of adhesion of the outer vertex of this segment. The free length of the radial segment decreases to a minimum free length during adhesion (becoming stiffer over time) [Equation 2.15]. After disruption the free length of the radial segment reverts back to a maximum free length [Equation 2.16].

Additionally, the alignment of the radial segments to the angle bisector is loosened during adhesion (like in the extended single chain model). This is also linked to the temporal change of the free length of the radial segments. The spring constant k_m of the torsion spring responsible for the alignment to the angle bisector is softened during adhesion (with the same rates like the change of the free length), in this model decreasing to zero and reverting back to full strength after disruption [Equation 2.7].

Contrary to the other models the torsion springs of the inner chain are not influenced by adhesion, the chain segments just tend to straighten for this chain.

The change of the free length L_0 of the radial segments is influencing the elastic forces F_e at the two neighbouring vertices [Equation 2.3], this is the main part of motorisation for this model, influencing the force equilibrium of the neighboured vertices.

2.3. Surface roughness

Surfaces with random roughness are one mean to get some insights into the behaviour and the capabilities of the three introduced models – as a method to test the reaction to obstacles.

Roughness is a measurement of the surface's texture. There are many roughness parameters to quantify the roughness quality of a surface. These parameters are derived from statistical analysis and signal processing, because a surface roughness can be mathematically considered as a spatially varying signal.

Amplitude parameters characterise the surface by the vertical deviations from the mean height. They are the most common parameters found in technical and engineering literature to describe the roughness of surfaces. One common parameter is the so called rms-height, the root mean squared average of the surface profile [26]:

$$R_q = \sqrt{\frac{1}{n}\sum_{i=1}^{n} y_i^2}$$

This parameter is giving a raw impression of the surface roughness but is lacking to characterise its "fidelity", because it is only dependent on the absolute deviation of the height from the mean height, giving no information on the spatial distribution of the height values. Thus, one profile parameter is not sufficient to describe all qualities of the roughness of a surface. The skewness and kurtosis of the spatial distribution of the height values are additional means to characterise surface roughness. Industrial standards take the average distance between the highest peak and lowest valley in a predefined sampling length to describe roughness. The mathematician Benoît Mandelbrot has shown that surface roughness is also connected with fractal dimension [28].

2.3.1. Modelling

Here, modelling of a random rough surface is based on the method by Garcia and Stoll (1984) [40]. It is only dependent on three parameters to define a random rough surface. The first parameter is the standard deviation σ of an uncorrelated Gaussian distribution of surface points $s_u(x)$ for each coordinate x with a discrete distance dx (the second parameter) along the surface, assumed as a one-dimensional height profile for simplicity (the standard deviation in combination with mean height of zero is corresponding to the previously introduced average height parameter, the rms-height, in the previous section). This uncorrelated Gaussian distribution $s_u(x)$ is then convolved with a Gaussian filter:

$$g(x) = \frac{1}{\sqrt{\pi}\frac{L_c}{2}} e^{\frac{-|x|^2}{L_c^2/2}}$$

$$s_c(x) = \int_{-\infty}^{+\infty} g(x-x') \cdot s_u(x') \, dx' \tag{2.17}$$

giving the third parameter L_c as the filter's correlation length of the convolved rough surface function $s_c(x)$. This convolution is most efficiently performed by using a discrete Fast Fourier Transform (FFT) algorithm. The uncorrelated surface function $s_u(x)$ is evaluated by Fast Fourier Transformation and is multiplied with the Fast Fourier Transform of the Gaussian filter and finally backtransformed by an inverse FFT to obtain $s_c(x)$ [40].

Recapitulating, the finally modelled random rough surface is defined and characterised by the standard deviation σ of the uncorrelated Gaussian distribution of surface points with a given distance dx between their x-coordinates and the correlation length L_c of the Gaussian filter.

Influence of modelling parameters on surface roughness

To get an impression on how the three different surface parameters influence the surface roughness, multiple plots of random surface profiles are generated with a set of standard parameters ($dx = 0.25$, $\sigma = 0.1$ and $L_c = 0.5$), where one of each parameter is screened with a range of values to show the change of surface profiles under influence of this parameter. These one-dimensional surface profiles are merged in a diagram with the surface

2. Modelling

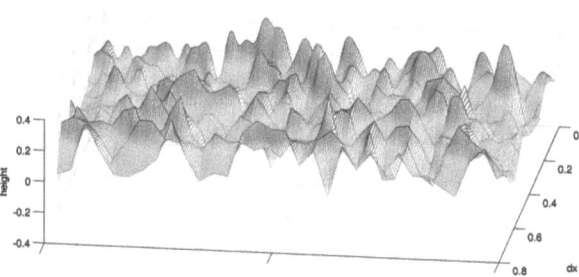

Figure 2.4.: The influence of the distance between surface points dx on surface roughness. An increase in dx results in more "edged" peaks.

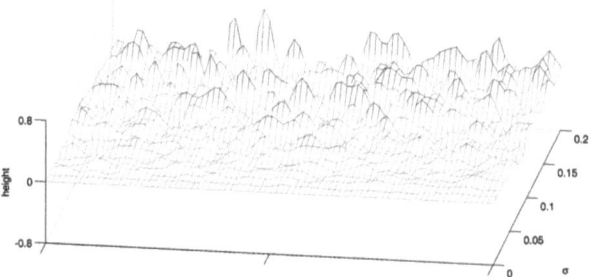

Figure 2.5.: The influence of the standard deviation σ on surface roughness. An increase in σ is increasing the overall roughness with randomly occuring higher peaks.

coordinates on the x-axis, the different one-dimensional surface profiles generated by the screening parameter on the y-axis, whereas the z-axis shows the height of surface roughness.

Figure 2.4 is showing the influence of the distance of surface points dx on the roughness of a surface: at first sight there does not appear any real change in the overall height profile, but at a closer look, an increase in dx is making the peaks more "edged", because there are less surface points smoothed by the Gaussian filter, whereas smaller distances produce more rounded peaks.

The influence of the standard deviation σ can be seen in Figure 2.5. As expected, the increase in σ is increasing the roughness of the surface, because an increase in the standard deviation causes higher peaks in the profile, and vice versa: for small σ, less roughness because of small peaks, and no roughness at all with a value of zero.

The influence of the correlation length L_c of the Gaussian filter on surface roughness

2.4. Parameter overview

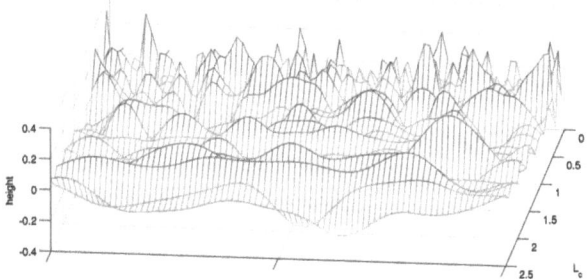

Figure 2.6.: The influence of the correlation length L_c on surface roughness. An increase in L_c is smoothing the peaks, reducing overall surface roughness.

is shown in Figure 2.6. As expected, the increase in L_c is reducing the roughness of the surface, because the peaks are smoothed, whereas a correlation length close to zero results in a merely unfiltered and non-smoothed surface.

2.3.2. Surface adaptation

The robot models need to adapt to surface roughness. It influences on how the outer segments are touching the surface and this is influencing the disruption of the rear vertex: The disruption is still controlled by Equation 2.12 for the extended single-chain model, where the perpetual direction is orthogonal to the mean of the surface, but for the other models, the disruption of the last attached vertex is dependent on the direction of the last resting segment (dependent on the grade of roughness it may not lying parallel to the mean surface), given by the formula:

$$\vec{F} \times \vec{d} \geq f_h \tag{2.18}$$

with $\vec{F}$ as the force vector pulling on the last attached vertex and $\vec{d}$ as the directional unit vector of the last resting segment and f_h the force height threshold as mentioned in Equation 2.12 since the length of the vertical force component need to pull on the vertex for disruption.

2.4. Parameter overview

The following tables are giving an overview about the parameters and non-dimensional values used in simulation for the models and the modelling of random rough surfaces:

- Table 2.1 contains the optimised parameter sets for each model.
- Table 2.2 contains the parameters used for modelling random rough surfaces.

2. Modelling

Figure 2.7 is giving an overview of examples of height profiles generated with the parameter values for surface modelling, which are used in simulations of the robot models on random rough surfaces. It shows an increase in roughness from the bottom left (small σ, large L_c) to the top right (large σ, small L_c).

Table 2.1.: The intrinsic model parameters and their non-dimensional values for the single-chain-model (s.-c.), extended single-chain-model (e. s.-c.) and double-chain-model (d.-c.). Value pairs in k_e and k_m are for the chain segments and the radial segments respectively. Value pairs in k_p and A_0 are for the area enclosed by the inner chain and the area for each peripheral chamber respectively. Missing values are not relevant for the corresponding model.

Symbol	Meaning	Non-Dimensional Value			Equation
		s.-c.	e. s.-c.	d.-c.	
N	number of vertices	24	20	16	
L_0	free length of segment springs		1		
k_e	spring constant for segments	5	20, 10	21, 16	2.3
k_p	pressure coefficient		0.1	0.1, 2.5	2.5
k_m	coefficient for moment of force	3	3.5, 8	1.5, 3	2.7
r_a	adaptation rate for adherent vertices	0.01	0.018	0.07	2.13, 2.15
r_f	adaptation rate for free vertices	0.0025	0.002	0.035	2.14, 2.16
L_{0min}	minimal adaptive radial free length		0.5		2.15
L_{0max}	maximal adaptive radial free length		1.25	1.5	2.16
f_h	vertical disruption limit	0.95	1.6	2.5	2.12
η	friction coefficient (simple friction)		2		2.9
η_u	friction coefficient (tangential)	1			2.10
η_n	friction coefficient (normal)	20			2.10
η_e	friction coefficient (elasticity)	1			2.10
η_m	friction coefficient (moment)	2			2.11
f_g	gravitational scaling factor		0.0025		2.2
A_0	equilibrium area	45.6	31.6	20.1, 0.8	2.5
dt	time step		0.05		2.1

Table 2.2.: The parameters used for modelling random rough surfaces and their non-dimensional range or values.

Symbol	Meaning	Non-Dimensional Value	Equation
σ	standard deviation of surface height	0.05 – 0.15 (in 0.025 steps)	
dx	distance between surface points $s_u(x)$	0.1, 0.25, 0.5	
L_c	correlation length of Gaussian filter	0.5 – 1.5 (in 0.25 steps)	2.17

2.4. Parameter overview

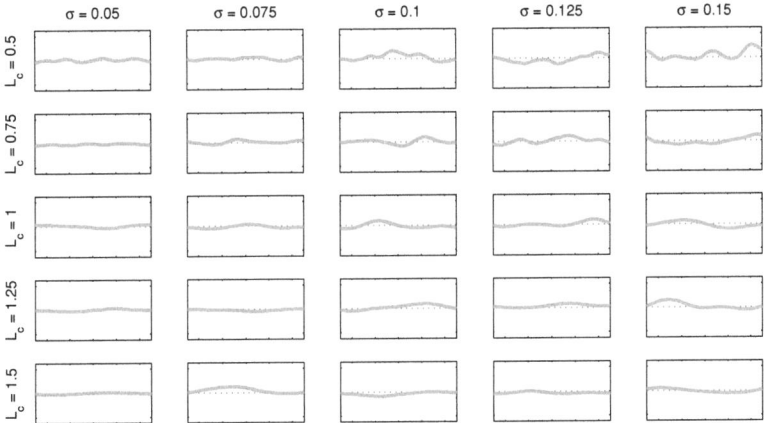

Figure 2.7.: Examples of surface height profiles with values of the standard deviation σ and the correlation length L_c given in the table of the rough surface parameters, which are used in simulations with random rough surfaces. Each axis tick represents a length of 1 unit. There is an increase in roughness from the bottom left to the top right.

3. Simulation results

Each of the three bio-inspired robot models has two stable dynamic configurations (depending on the given set of parameters) – the first is a stationary configuration in which the model does not perform any persistent locomotion but both ends of the model are steadily fluctuating by detaching and reattaching to the surface – the second and more interesting configuration is a dynamic behaviour of persistent locomotion where the model moves along the surface in one direction with a certain averaged translocation speed that is maximal for an optimised set of parameters. Depending on circumstances the models can switch between these configurations, e. g. due to obstacles, such as surfaces with strong roughness, or by changing some of the intrinsic model parameters. To analyse the models' overall performance in configurations with persistent locomotion and in order to construct a potentially robust bionic robot locomotion device it is of main interest that all parameters are optimised for the best translocation speed. Moreover, with the help of statistics it is possible to characterise the models' behaviour by measuring certain indices of performance and shape deformation.

3.1. Overall performance

Figure 3.1 depicts a snapshot of the three models during a persistent movement phase – observable is the model's shape and its deformation compared to the starting configuration [Figure 2.1]. The front end is defined as the side of the model which is pointing to the direction of movement (in this case the right side). The data used for analysis are taken either from one full locomotion cycle of each model – meaning the time span from the disruption of one vertex over reattachment until its next disruption – or from a fraction of this locomotion cycle, in numbers (for standard parameters [Table 2.1]: time of one cycle in simulation on a flat non-rough surface for the single-chain model is 7640.4 time steps, for the extended single-chain model 2727.2 time steps and for the double-chain model 256.3 time steps. During movement the general observation is that the singe-chain models are more elongated compared to the starting configuration (and compared to the double-chain model during movement), demonstrating a persistent creeping or crawling locomotion, whereas the double-chain model is nearly circular in its outer shape and demonstrates a persistent rolling locomotion, revealing a noticeable shear of the inner chain relative to the outer chain with the connecting radial segmental spokes always inclined to one side. Thus, a vertex on the inner chain is always ahead of the same vertex of the outer chain during movement.

3. Simulation results

(a) single-chain model: creeping locomotion

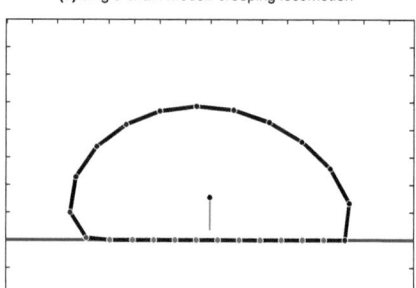

(b) extended single-chain model: crawling locomotion

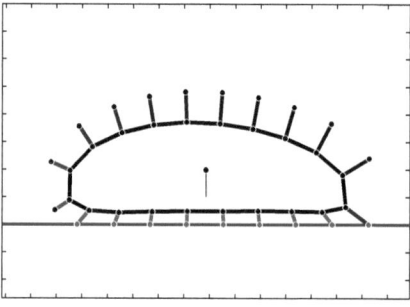

(c) double-chain model: rolling locomotion

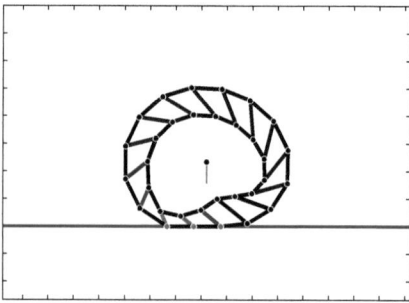

Figure 3.1.: The persistent movement from left to right with distinctive shape deformations. The vector at the centre point is the gravity vector, showing the direction and magnitude of gravity. Shading of grey defines the elasticity gradient.

3.1.1. Translocation speed

The translocation speed of each model's persistent locomotion is the statistical mean of the displacement per time step of the model's centre point parallel to the surface. Figure 3.2 compares the mean translocation speed of each model and reveals some distinctions. The double-chain model has the highest mean speed (~6.59×10^{-2} length units per time step) – due to its rolling locomotion – but significantly slows down for a short time after each vertex disruption [Figure 3.2c]. The disruption is interfering with a consistent rolling locomotion, therefore detachment of rear vertices is a brake for rolling locomotion. The extended single-chain model is moderately slower (~1.05×10^{-2} length units per time step), because it is rather crawling (using its spikes as feet) instead of rolling, but increases significantly in speed for a short time after each vertex detachment [Figure 3.2b]. Each detachment at the rear end is substantially pushing the model forward (as shown by the high peaks shortly after detachment), which allows the front end to initiate a new attachment, therefore detachment of rear vertices is an accelerator for crawling locomotion. The single-chain model is the slowest moving model (~2.62×10^{-3} length units per time step) with observable creeping locomotion. The translocation speed stays relatively on the same level without large divergences after detachment as seen in the other two models [Figure 3.2a], but the slight increase after disruption attests similarities to crawling locomotion.

Taking a closer look at one period of the locomotion cycle (the time interval between the detachment of single vertices) reveals small saltuses of the graph of translocation speed. These saltuses are caused by new attachments of vertices at the front end. In the extended single chain model, one saltus is visible close to a new detachment. In the single chain model there is a small oscillation close to the middle of the period.

As seen by the shape deformation during movement [Figure 3.1], the single chain models are both relatively elongated along the surface with many vertices attached due to the creeping and crawling locomotion, while the double chain model is more or less circular with only a few attached vertices due to the rolling locomotion with the fastest mean translocation speed. This is comparable to observations of living migrating cells: cells with strong attachment are moving considerably slower than cells with weak attachment (e. g. the nearly immobile neurons of the sea slug *Aplysia* on sticky substratum in culture, in contrast to the very fast and weakly adhesive fish keratocytes as two extreme examples [4]).

In summary, a rolling locomotion is on average faster than a crawling locomotion which is still significantly faster than the creeping locomotion. The required disruption of the adherent vertices at the rear end is negatively influencing the rolling locomotion but is reinforcing the crawling locomotion.

3. Simulation results

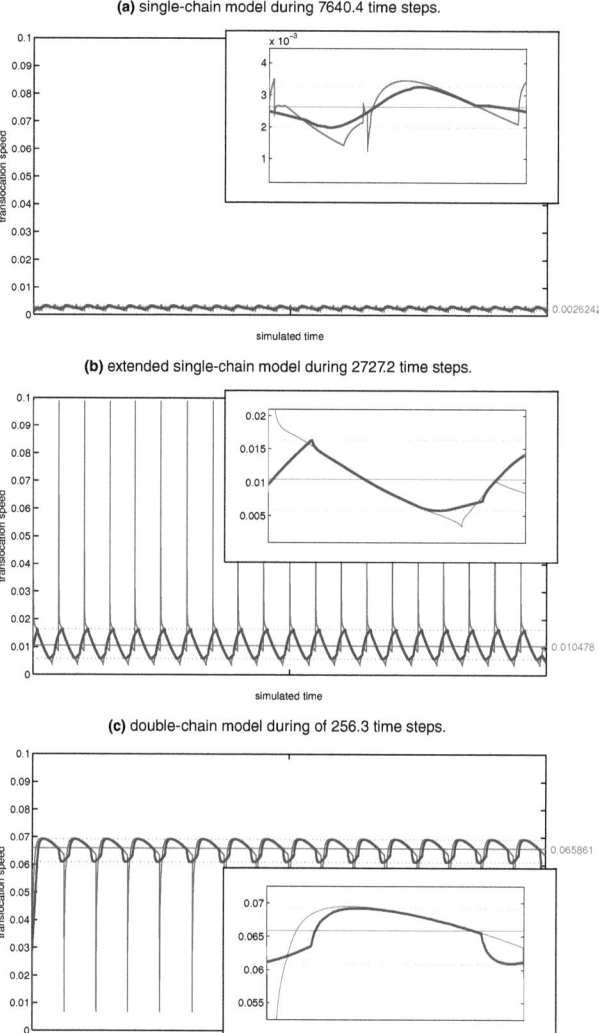

Figure 3.2.: The translocation speed (grey) of the models. The **bold dark graph** is a smoothed version with a moving average of one third of the period length. The grey line marks the mean of the smoothed graph and the dotted line its maximum and minimum. Data is shown for the time of one full locomotion cycle. Each inset depicts one period between the detachment of vertices.

3.1.2. Adhesiveness

The quantification of the models' adhesiveness is related to the average height (or the mean distance of all vertices to the surface) of a model, because the more vertices are attached to the surface the lower is the average height. Besides, it also reveals a first insight into the temporal change of shape of the model – the lower the average distance to the ground surface becomes, the flatter and the more elongated is the robot model along the surface. Figure 3.3 is showing the temporal course of the average height for each model during one quarter locomotion cycle. Comparison the three models reveals that the extended single-chain model has much more variation in the height during movement [Figure 3.3b], the double-chain model lies between [Figure 3.3c], whereas the single-chain model has the least variation and shows a sub-periodicity [Figure 3.3a]. These saltuses are again located close to the middle of a period in the single-chain model and close to a new detachment in the extended single-chain model (cf. translocation speed analysis). Thus, a new attachment of the frontal vertex is causing the sub-periodicity of the single-chain-model.

3.1.3. Polarity

Though the average height gives an impression of the general alteration of shape, there is no information of differences in details, e. g. between front and rear profiles, particular an asymmetry in shape caused by different deformations at the front and rear end. This asymmetric deformation of the model defines its polarity. This polarity is quantified by statistical means: Each chain segment has a certain deflection to the next segment, and the deflection angle at the vertices is roughly describing the local chain curvature, which is a more detailed characterisation of shape differences, respectively the polarity. Plotting the segment angles at the free unbound vertices of the (outer) chain from the rear to the front end results in a simplified curvature graph seen in Figure 3.4.

Comparing the perfectly symmetric starting configuration [Figure 3.4a] with the situation during locomotion [Figure 3.4b] shows a significant symmetric shift of this curvature graph. It has become left-tailed – the left side (corresponding to the rear end of the model) is more elongated and the curvature peak is closer to the right side (corresponding to the front end of the model), whereas in the graph of the starting configuration the curvature peak is exactly located in the middle and both tails have the same type of local maximum at the adhesion vertices. Moreover, since the left tail also has lower values than the shorter and steeper right tail [Figure 3.4b], this corresponds to a geometry of the robot model where the shape of the rear end is less curved and thus steeper contrasting with a more curved and thus flattened front end. A simple mathematical method to characterise the asymmetry of the shape of a distribution is skewness.

3. Simulation results

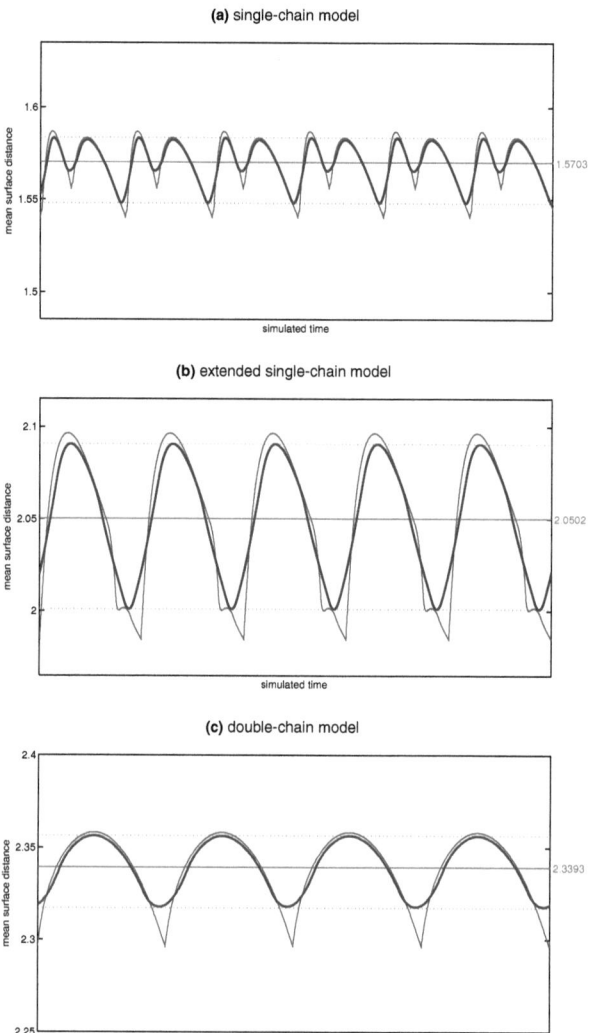

Figure 3.3.: The adhesiveness as a measurement of the mean distance (grey) to the surface. The **bold dark graph** is a smoothed version with a moving average of one third of the period length. The grey line marks the mean of the smoothed graph and the dotted line its maximum and minimum. Data is shown for the time of one quarter locomotion cycle.

3.1. Overall performance

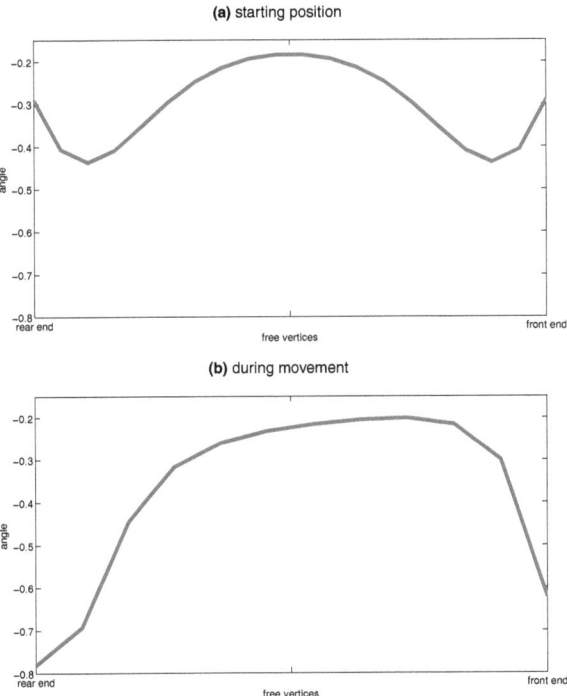

Figure 3.4.: The chain curvature graph as the deflection angles of the free segments at the starting position [Figure 2.1a] with even tails (and the same type of of local maxima at the adhesion node) and the left-tailed situation during locomotion [Figure 3.1a] (both for the single-chain-model). The skewness of the curvature graph is a measurement of the asymmetry in model's shape and quantifies the polarity of a model during locomotion.

The skewness of a distribution with variable x is defined as third standardised moment:

$$\gamma = \frac{E(x-\mu)^3}{\sigma^3}$$

where μ is the mean of x, σ is the standard deviation of x and E is the expectation operator. This non-dimensional value describes the mentioned asymmetry, because a negative skew characterises a left-tailed distribution and a positive skew a right-tailed distribution, whereas a skew of zero means perfectly symmetric sides. Additionally, the magnitude of the value of skewness characterises the degree of asymmetry – a value close to zero means only a slight asymmetry. Here, the polarity of the models is simply defined as the negative skewness of the curvature graph.

3. Simulation results

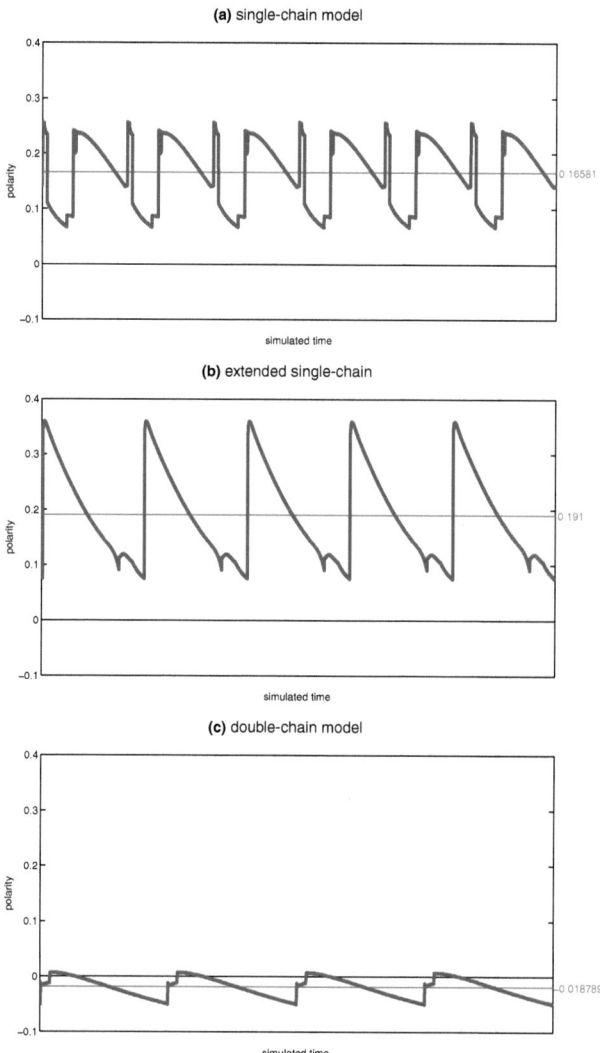

Figure 3.5.: The temporal change of the models' polarity during one quarter locomotion cycle. The grey line marks the mean polarity value.

Figure 3.5 shows the temporal change of polarity of the models during one quarter locomotion cycle. The extended single-chain model has the highest mean polarity value [Figure 3.5b], followed by the single-chain model [Figure 3.5a] (again with visible subperiodicity), whereas the double-chain model's mean polarity is close to zero but slightly negative [Figure 3.5c]. The visible saltuses of one period indicate again the time a new vertex attaches to the surface at the front end.

Since the double-chain model demonstrates a rolling movement on the surface, the slightly negative mean polarity indicates the following detail: after each vertex detachment the polarity becomes slightly positive, thus shortly after disruption the rear end is less curved which changes temporally and gradually to become more curved while rolling on the surface. But this observation is only valid for the outer chain, the inner chain has a continuingly positive polarity [Figure 3.1c], which is neutralised by the deformation of the outer chain. The extended single-chain model is the most polarised model during its persistent crawling locomotion. It has the highest mean polarity value of all models, indicating a less curved rear end and a more curved (and flattened) front end, which fits to the crawling movement. Additionally, it has high peaks of polarity after each vertex detachment. The simple single-chain model has the same asymmetry, but with a slightly weaker magnitude in polarity than the extended single-chain model.

The general observation for all models: the detachment of rear vertices during locomotion is causing a short temporal increase in polarisation – in other words: detachment of the rear end enhances polarisation.

3.1.4. Forces

During movement a force is pulling on each vertex causing the displacement of the free and unbound vertices. Figure 3.6 depicts the extended single-chain and the double-chain model with a situation close before the disruption of the rear vertex with plotted force vectors acting on the vertices. The double-chain model has a large force pulling at the last attached vertex against the direction of movement and only a small force at the front, this strong force might be a result of the shear of the chains. The extended single-chain model has strong forces at the attached front vertices pulling against the direction of movement, whereas at the attached rear vertices the force vectors are pulling in direction of movement. The sum of the horizontal force components at the front attached vertices is bigger than the sum of the rear horizontal components. This is important, because for the attached vertices a negative horizontal force component (negative means pointing against direction of movement) is in a simplified view the traction force responsible for locomotion. A positive horizontal force is acting as a rearward traction force or as braking force, which is on a first thought counterproductive in the sense of applying a net forward force purely by applying traction forces, but it may also be advantageous for the translocation, especially for retraction. The positive vertical force component is the potential disruption force, responsible for detachment of vertices [subsection 2.2.3] (whereas a neg-

3. Simulation results

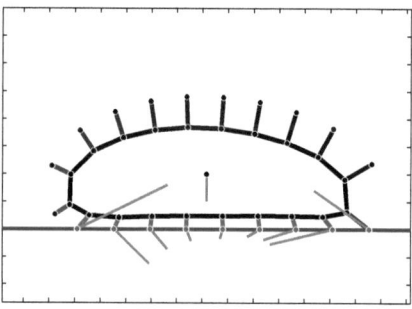

(a) extended single-chain model: crawling movement

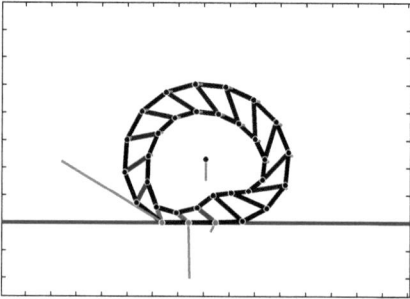

(b) double-chain model: rolling movement

Figure 3.6.: The force vectors acting at each vertex during movement. The movement direction is from left to right. The strongest forces are acting on the attached vertices. The vector at the centre point is the gravity vector, showing the direction and magnitude of gravity. Shades of red colouring define the elasticity gradient.

ative vertical force component amplifies adhesion by pressing the vertex onto the surface).

Thus for disruption, looking at the vertical force component of the first adherent vertex at the front and the last adherent vertex at the rear end is another possibility to characterise the models' asymmetry and behaviour [Figure 3.7]. The vertical force component at the frontal vertex of the double-chain model is very small [Figure 3.7c]. In contrast, the vertical force component of the first adherent vertex of both single-chain models is increasing rapidly after attachment, reaching a small peak close to the disruption limit as seen for the extended single-chain model [Figure 3.7b]. For the simple single-chain model it even reaches the disruption limit [Figure 3.7a], causing a short disruption of the frontal vertex (which reattaches soon after), which is an explanation for the observed sub-

3.1. *Overall performance*

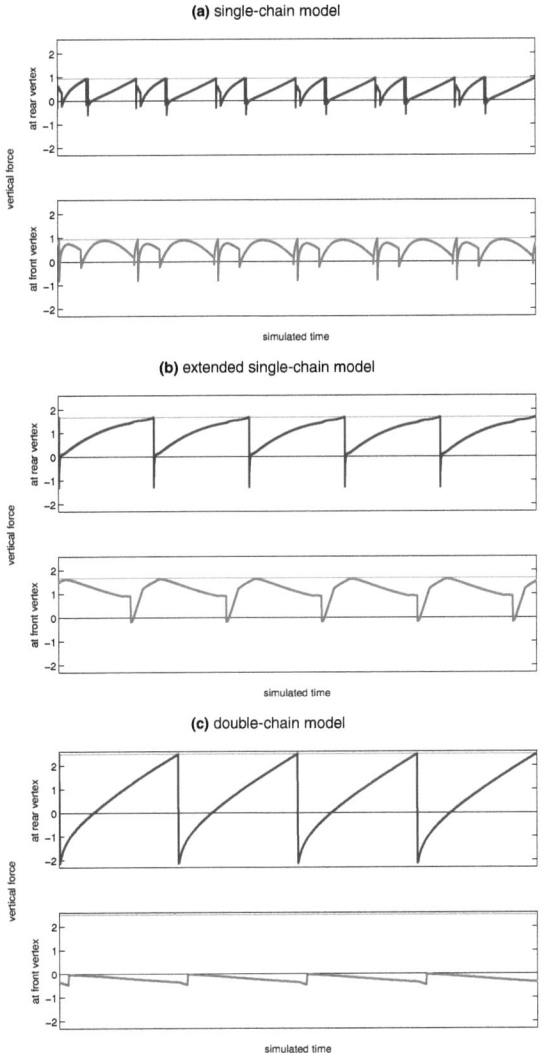

Figure 3.7.: The disruption force (as the vertical component of force acting on the vertex) of the first attached front and last attached rear vertex. The grey line marks the limit when the vertex is disrupted. Data is shown for one quarter locomotion cycle.

51

periodicity: the front vertex attaches but prematurely detaches soon after, causing a slow down and reattaches again.

Recapitulating, the creeping and crawling locomotion is causing a strong vertical force component for freshly attached vertices, whereas the rolling locomotion is causing only a small vertical force component. The slow creeping locomotion (and the sub-periodicity) is caused by a short temporal detachment of a newly attached front vertex.

The analysis of the horizontal force components reveals another distinctive difference between the models and the different types of locomotion [Figure 3.8]. In the double-chain model the traction force is mainly caused by the rear end [Figure 3.8c]. This explains why the speed significantly falls after detachment in the double-chain model [Figure 3.2c], because with the disruption of the last vertex the model loses its main contributor of traction: The model is losing "grip". Additionally, the strong negative horizontal force component will also act on the freshly detached vertex, pulling it against the direction of movement, thus also decelerating the model. This is different compared to the extended single-chain model [Figure 3.8b]. In this model the front vertices are the main contributor to the traction force, whereas the forces acting at the rear end are a braking force. This seems to be a disadvantage, but considering that the vertices of the rear end will eventually getting disrupted, this positive force component is then pulling the new free vertices in direction of movement, meaning the retraction (caused by the detachment of rear vertices) is pushing the model forward. This explains the short speed boost after each detachment in the extended single-chain model [Figure 3.2b]. Considering the simple single-chain model the analysis of horizontal force components again shows the sub-periodicity, but with the main difference, that both front and rear vertices contribute traction forces (but only with a low absolute value compared to the other models), though there is a short period, when the rear vertex has a positive horizontal force component [Figure 3.8a].

The double-chain model and the extended single-chain model are selected for a more detailed insight into the generation of traction and retraction forces during adhesion, because they demonstrate the best performance for rolling and crawling movement. Figure 3.9 shows the temporal development of the horizontal and vertical force components acting on one vertex during its time of attachment. In the double-chain model only a weak traction force acts early on after adhesion, defined by a small negative horizontal component. [Figure 3.9b]. Its sudden drop-down after two thirds of adhesion time corresponds to the time of detachment of the previous attached vertex (the last third of the graphs corresponds to each period seen in Figure 3.8c). Only after this time a large traction force is generated. This confirms the previous statement that the last adherent vertex is the main contributor to the overall traction force and the detachment of this vertex is causing the significant temporal slow-down seen in the translocation speed analysis [Figure 3.2c]. The disruption force (the positive vertical component) causing this detachment is also generated only when the vertex becomes the final attached node at the rear end.

In the extended single-chain model the negative horizontal force component, respect-

3.1. *Overall performance*

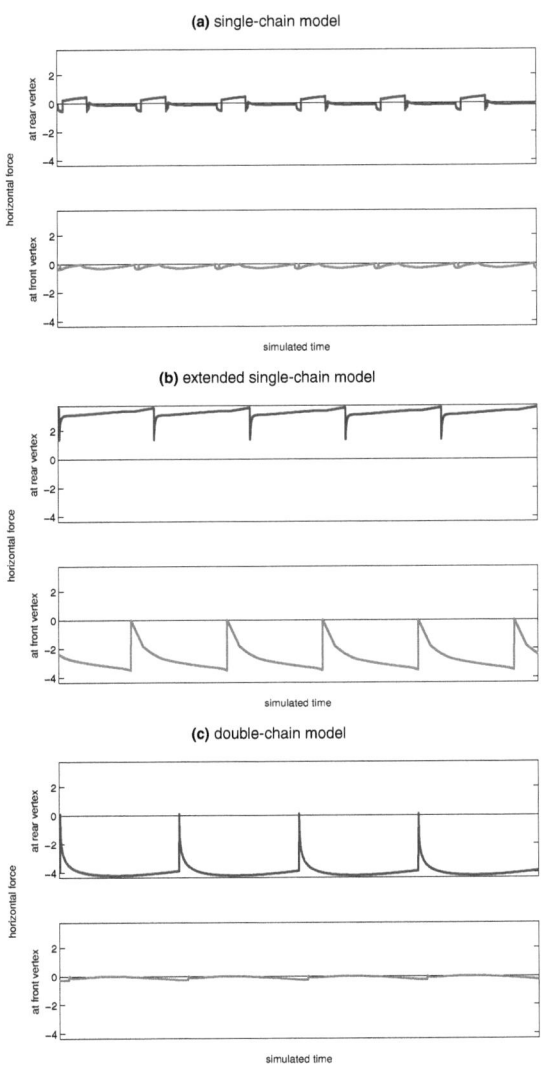

Figure 3.8.: The traction force (as the horizontal component of the force acting on the vertex) of the first attached front and last attached rear vertex during the time of adhesion of this vertex. Data is shown for one quarter locomotion cycle.

3. Simulation results

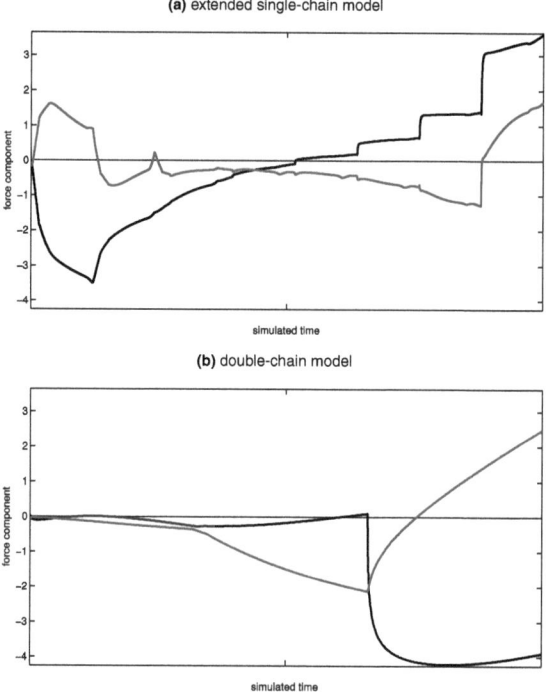

Figure 3.9.: The **horizontal** and vertical force components of one vertex during the time of its adhesion. A positive vertical component is the disruption force pulling on the vertex that eventually detaches the vertex. A negative horizontal force is corresponding to a traction force, whereas the positive value can be considered as a braking force.

ively the traction force, is mainly generated at the beginning of adhesion [Figure 3.9a].

After half the adhesion time the negative horizontal force component becomes positive, resulting in a braking force. Comparing this with the geometry of the extended single-chain model [Figure 3.1b] this is the time when the spike of the adherent node is changing its alignment. At the front, when a new vertex attaches, the alignment of the spike is to the right side into direction of locomotion, whereas on the rear it is aligned to the left side against the direction of locomotion. The attached vertices and the corresponding elastic (and stiff) spikes are becoming anchors over time which hinder the forward movement. Therefore, as confirmation of the previous assertion, the speed is temporally boosted after each disruption [Figure 3.2b], because it loses a hindering anchor. The increase in the positive horizontal force component occurs step-wise, each small saltus of increase is caused by a detachment of a previous vertex, but the largest saltus (and the increase in braking force) happens after the detachment of the last of the previously attached vertices. This is the point in adhesion time, when the disruption force becomes large enough for disruption (but there is also a positive vertical force component at the beginning of adhesio which is getting close to the limit required for disruption). The final part after the last saltus is corresponding to the previous force analysis at the rear end [Figure 3.7b, Figure 3.8b].

In summary, there are distinctive differences in traction and retraction forces of the rolling locomotion of the double-chain model and the crawling locomotion of the extended single-chain model. The traction of rolling locomotion is mainly caused by the rear end but it is counterproductive for retraction after detachment (temporal slow-down of translocation speed). In contrast, the traction of crawling locomotion is mainly caused by the front end and becomes a braking force at the rear end. But this force is advantageous for retraction (temporal boost of translocation speed). For comparison, see Figure 3.6 with the strongest forces acting on the attached vertices during locomotion.

3.1.5. Mechanical stress

Working forces on vertices also cause mechanical stress within the chain segments. For a characterisation of the involved mechanics at front and rear end of the models, the examination of the mechanical stress of neighbouring segments of the first and last attached vertex is performed. The mean mechanical stress of four segments neighbouring the first and also the last attached vertex is measured (in detail: at each vertex the two closest attached segments and the two closest free segments are considered for stress calculation). A negative stress value generally means compressed segments, whereas a positive stress value generally means that the segments are expanded. A value of zero indicates completely relaxed segments. The mean mechanical stress of the segments at the front and rear end is shown in Figure 3.10.

In general, the rear segments of the single-chain model are compressed, whereas the front segments are expanded [Figure 3.10a]. The front segments are stressed the most by changing from a strong compression to a strong expansion, changing back to a compressed state and again return to an expanded state. This stress is caused by the short detachment and reattachment of the front vertex, as discussed earlier. The strong drop of the graph from expansion to compression is caused by the attachment of a new vertex, the increase to expansion shortly after this time corresponds to the interval, when the front vertex is temporally detached, while the second drop to compression is the reattachment, when the vertex stays attached. This change from compression to expansion and back follows the sub-periodicity caused by this event. In contrast the chain segments of the extended singe-chain model are relatively relaxed and only slightly compressed, but the rear segments experience a strong compression after the disruption of the last vertex and the front segments a slight compression shortly after a new adhesion [Figure 3.10b]. The

3. Simulation results

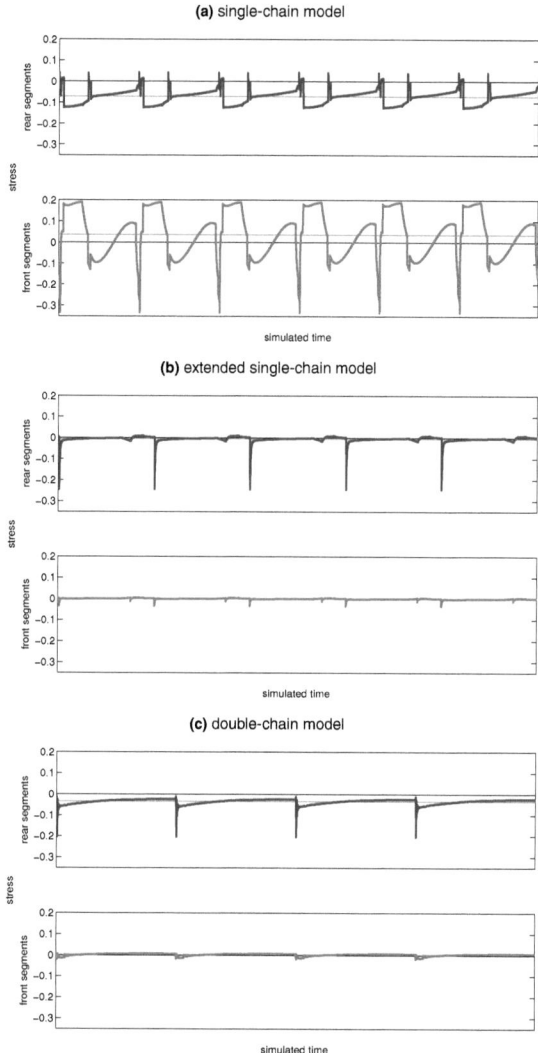

Figure 3.10.: The mechanical stress of segments at the rear and the front end. Negative values indicate compressed segments, while positive values indicate that the segments are generally expanded. Data is shown for one quarter locomotion cycle.

same mechanics is observable for the front segments in the double-chain-model, but the rear segments are always compressed in this model and getting more compressed after disruption for a short moment Figure 3.10c].

In summary the mechanical stress of the chain segments at front and rear is quite different in each model. In the single-chain-model the segments experience the most stress in absolute values. In the other models the segments mainly experience compression, especially after disruption, while the double-chain model's rear segments are generally compressed, whereas the rear segments of the extended single-chain model are only slightly compressed.

3.1.6. Correlations

The translocation speed of a model will dynamically define its shape and the shape will dynamically define the translocation speed. Cross-correlations between the smoothed temporal course of translocation speed [subsection 3.1.1, Figure 3.2], the smoothed temporal course of the mean height [subsection 3.1.2, Figure 3.3] and the temporal course of polarity [subsection 3.1.3, Figure 3.5] might help to quantify this relationship between translocation speed and shape during movement. This correlation analysis is concentrating on the extended single-chain model and the double-chain model as the best performers in translocation speed and their distinctive difference in crawling and rolling locomotion.

Figure 3.11 shows the cross-correlation between the smoothed translocation speed and the mean height as seen in Figure 3.2 and Figure 3.3. It reveals that there is a close relationship between translocation speed and height in the double chain model with a correlation time of zero [Figure 3.11b]. In the extended single-chain model there is a small temporal lag between the maximum of speed and the maximum of the mean surface distance [Figure 3.11a], but with a strong correlation.

Figure 3.12 shows the cross-correlation between the smoothed translocation speed and the polarity as seen in Figure 3.2 and Figure 3.5. There is a time lag between translocation speed and polarity in both models. Due to the strong correlation it is difficult to quantify, if polarity is leading the translocation speed or vice versa.

3.2. Parameter screening

The performance (especially in translocation speed and polarity) of the models is depending on their intrinsic properties, which are defined by the model parameters. All the dynamics in the models is based and optimised on dynamically altering the elastic properties and creating an elasticity gradient. The model's performance is basically influenced by the static elasticity of their segments, the bending stiffness of the chains [section 2.2.1] and the adaptation rate responsible for the elasticity gradient [subsection 2.2.3]. The corresponding parameters that determine these elastic properties are the elasticity parameter k_e

3. Simulation results

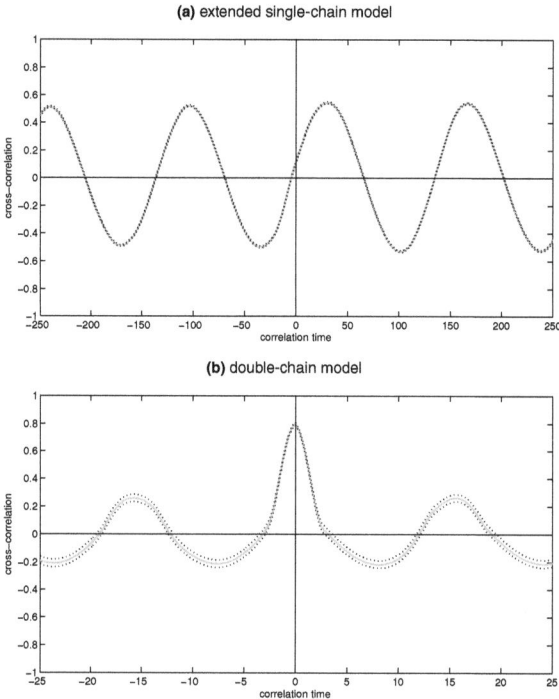

Figure 3.11.: The cross-correlation between the smoothed temporal course of translocation speed [Figure 3.2] and mean height [Figure 3.3] for the extended single-chain model and the double chain model. The extended single-chain model has its maximum of correlation at a correlation time of ~30 time steps, whereas the double-chain model has a correlation time of zero. Black dotted lines define the confidence interval of correlation.

for the elastic segments [Equation 2.4], the parameter responsible for the strength of the bending stiffness of the chain k_m [Equation 2.7] and the rate r for the temporal change of elasticity [Equation 2.13, Equation 2.13, Equation 2.15, Equation 2.16].

A screening of this parameters by decreasing or increasing their value reveals on how these parameters influence the performance and the movement behaviour. For the screening the translocation speed and polarity are measured over a given constant time interval for each model after changing the parameter value, which might cause visible fluctuations of the data. A measurement over the time of one movement cycle has the disadvantage of varying time intervals, because theses are influenced by the translocation speed of one movement cycle. For data independent of speed a fixed time interval was chosen.

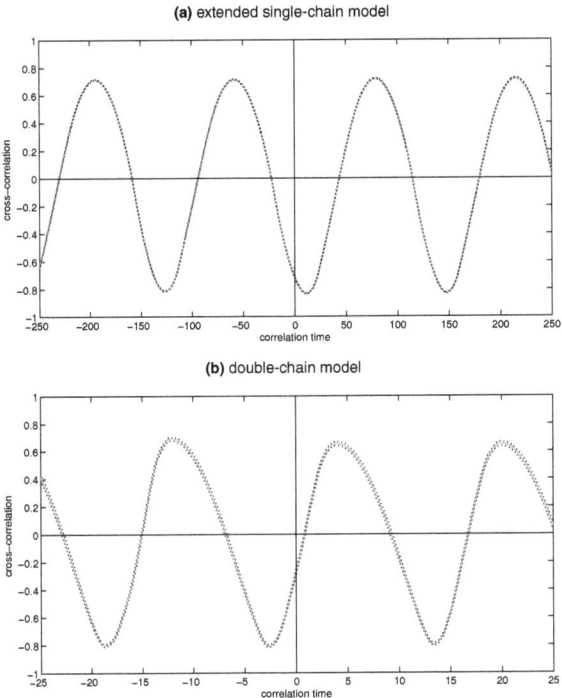

Figure 3.12.: The cross-correlation between the smoothed temporal course of translocation speed [Figure 3.2] and polarity [Figure 3.5] for the extended single-chain model and the double chain model, depicting the temporal lag between the two. Black dotted lines define the confidence interval of correlation.

3.2.1. Elasticity

Figure 3.13 presents the parameters screening of the elasticity parameter k_e in relation to the resulting speed and polarity of each model. Data is shown only for the parameter range, where a model performs a stable persistent locomotion. The single-chain model has only a small parameter range around the optimal value, where it is operating with a persistent locomotion. The extended single-chain model works in a broader range with decreased parameter values. The double-chain model is able to perform in a very broad range. In detail, the influence of k_e to the translocation speed is (as Figure 3.13a shows): In the double-chain-model a lower value of k_e results in a lower mean speed. 50% of the initial k_e value results in roughly 74% of the initial mean speed and is slowly converging to an optimum in translocation speed by increasing k_e beyond the standard value. Increasing the parameter beyond the maximum of translocation speed results in a faster decrease

3. Simulation results

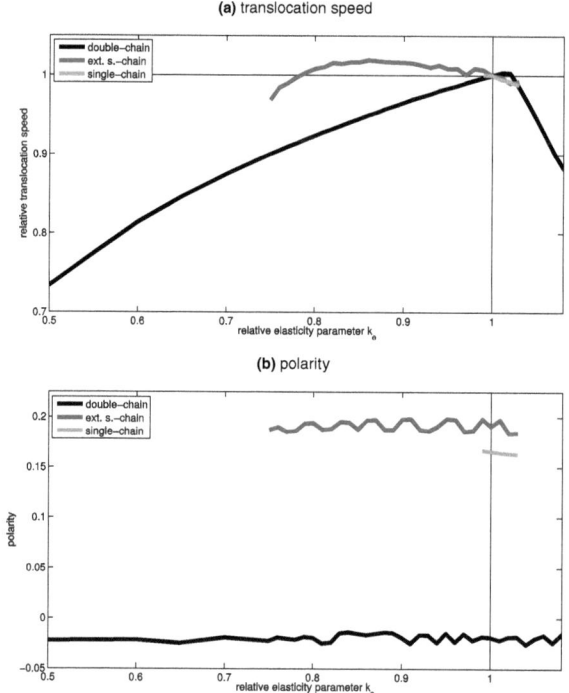

Figure 3.13.: Screening of elasticity parameter k_e relative to the optimised value used in simulations. Translocation speed is relative to the standard mean speed [Figure 3.2]. Data is showing only the range with a stable persistent movement.

of speed: too strong elastic segments is disadvantageous for locomotion. The extended single-chain has a flat optimum at approximately 85% of the standard value. The single-chain model requires an exactly adjusted elasticity parameter with such a small operating range.

Considering the influence on shape, Figure 3.13b shows that k_e has no direct impact on polarity. Except of some fluctuations, the polarity stays more or less on the same level in the operating parameter range. A small decrease in polarity is only visible for the single-chain model.

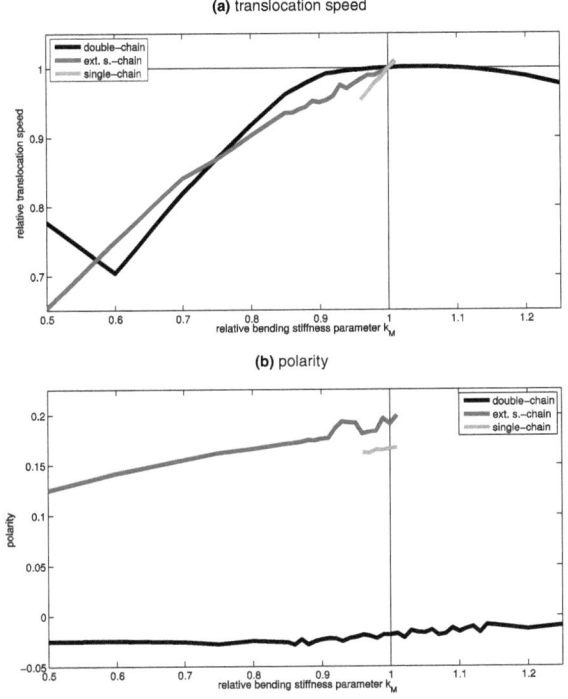

Figure 3.14.: Screening of parameter k_m defining the bending stiffness of the chain relative to the optimised value used in simulations. Translocation speed is relative to the standard mean speed [Figure 3.2]. Data is showing only the range with a stable persistent movement.

3.2.2. Bending stiffness

The model's dependency of the strength of the bending stiffness is shown in Figure 3.14. The single-chain model has only a small range for k_m with a persistent locomotion, whereas the extended single-chain model is also operational with lower values of k_m. The double-chain performs on low and high values with a very broad translocation speed optimum at approximately 105% of the standard value [Figure 3.14a]. The speed graph of the double-chain model demonstrates an asymptotic convergence to its optimum, increasing asymptotic from the lower parameter range and decreasing after the optimum. The translocation speed of the extended single-chain model is also increasing from lower parameter values with a slight asymptotic curvature, but it becomes unstable with high k_m values. The simple single-chain model also performs an increase in speed with increase in the bending stiffness strength in its narrow operating range.

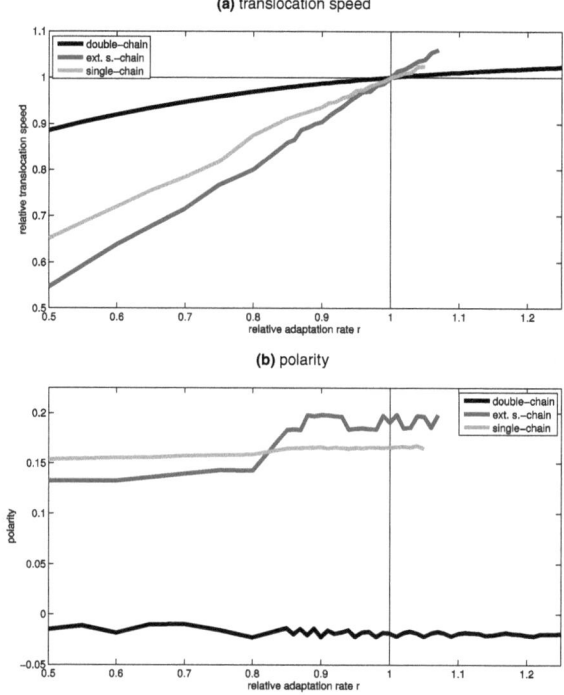

Figure 3.15.: Screening of the adaptation rate r defining the temporal elasticity change of the chain relative to the optimised value used in simulations. Translocation speed is relative to the standard mean speed [Figure 3.2]. Data is showing only the range with a stable persistent movement.

The bending stiffness parameter k_m has an impact on shape, respectively the polarity of the models [Figure 3.14b]. All three models show an increase in polarity with increasing k_m (more precisely for the double-chain model: the negative polarity value is getting closer to zero). The extended single-chain model has the steepest slope of increase.

In summary, the strength of the bending stiffness of the chains has an impact on polarity, a stronger bending stiffness is increasing the polarity. This is expectable, since the bending stiffness is defining how rigid the chain behaves on its deformations.

3.2.3. Elasticity adaptation

The temporal elasticity gradient is the motorisation of the models. This gradient is determined by the adaptation rates r for the adherent and non-adherent status [subsection 2.2.3]. These rates should have an impact on translocation speed. Figure 3.15 clearly depicts this. The three models substantially perform slower with lower adaptation rates [Figure 3.15a]. Both single-chain models show a linear increase in translocation speed with an increase in r, but they are limited to approximately 105% (single-chain model) and 107% (extended single-chain model) value of the standard rate before unstable behaviour occurs, which breaks the persistent movement. The double-chain model shows an slightly asymptotic increase in translocation speed and has still some room for improvement with much higher adaptation rates.

In case of polarity, the adaptation rate has no impact on polarity of the double-chain model, but the polarity of the single-chain models increase with increasing adaptation rates. At lower values the simple single-chain model is more polarised than the extended single-chain model, after around 80% of the standard rates they change their positions and the extended one becomes the most polarised model.

3.3. Rough surface performance

Rough surfaces are some kind of obstacles for the locomotion of the models, which they need to encounter with their adaptive flexibility in shape. Figure 3.16 shows the locomotion of the extended single-chain model and the double-chain model on a rough surface. The extended single-chain model should have less problems with strong roughness, because it should be able to step with its spikes over roughness peaks, "floating" over the roughness [Figure 3.16a]. The double-chain model has to roll over any unevenness which should have an influence on translocation speed [Figure 3.16b]. To test this hypothesis, the data of multiple simulation runs with both models moving on random rough surfaces (which were generated with the parameters in Table 2.2) was analysed in relation to the influence of the roughness on translocation speed and general shape. The parameter dx, defining the distance between surface points, showed no significant differences and is therefore omitted from the analysis.

3.3.1. Translocation speed

Figure 3.17 depicts the difference in translocation speed (compared to the mean translocation speed in Figure 3.2 on even surface) in relation to increasing roughness for the extended single-chain model and the double-chain model. The increasing roughness is corresponding to Figure 2.7 from bottom left to top-right, generated with the according standard deviation and correlation length. For characterisation of the roughness, the rms-height value R_q [section 2.3] is used. In this case the rms-height approximately corresponds to the standard deviation used for surface generation. Data is shown together with an estimated 95% confidence interval of the mean. The analysis reveals that the double-chain performs better in the case of translocation speed with a low rough-

3. Simulation results

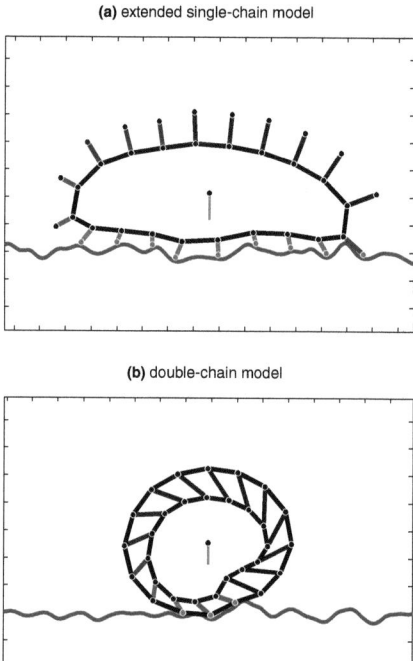

Figure 3.16.: The models performing on a random rough surface with standard deviation $\sigma = 0.15$ and correlation length of the Gaussian filter $L_c = 0.5$. The movement direction is from left to right. The vector at the centre point is the gravity vector, showing the direction and magnitude of gravity. Shades of red colouring define the elasticity gradient.

ness compared to an even surface, but on a rougher surface, the double-chain model is significantly slower. The extended single-chain model is always slower compared to an even surface, but a higher roughness has less impact on translocation speed. The extended single-chain model is performing better on higher roughness than the double-chain model, which approves the previous hypothesis.

3.3.2. Adhesiveness

Figure 3.18 shows the difference in mean height (compared to the mean height in Figure 3.3 on even surfaces) in relation to increasing roughness for the extended single-chain model and the double-chain model. The same surface profiles as in the translocation

3.3. Rough surface performance

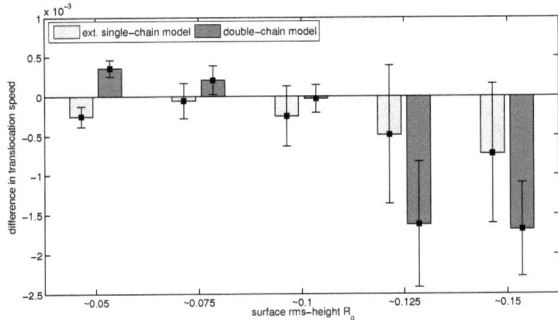

Figure 3.17.: The difference in translocation speed on rough surfaces for the extended single-chain model and the double-chain model (in relation to the mean translocation speed on plain surface [Figure 3.2]) with an estimated 95% confidence interval of the mean. The increasing surface roughness is corresponding to Figure 2.7 from bottom left to top right generated with the according standard deviation and correlation length. The extended single-chain model is better in adapting to surfaces with higher roughness compared to the double-chain model.

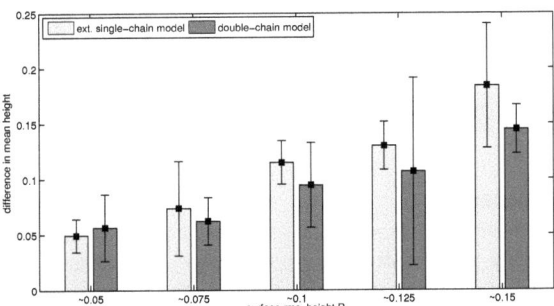

Figure 3.18.: The difference in the mean height on rough surfaces for the extended single-chain model and the double-chain model (in relation to the mean height on plain surface [Figure 3.2]) with an estimated 95% confidence interval of the mean. The increasing surface roughness is corresponding to Figure 2.7 from bottom left to top right generated with the according standard deviation and correlation length. A general increase in mean height is explainable, because of the increase in the rms-height of the rough surface, but the extended single-chain has a higher increase in the mean height compared to the double-chain model.

speed analysis were used. The increase in the mean height for increasing roughness is explainable due to the increase of the overall rms-height of the surfaces, though statistically there is no significant influence of roughness on the general shape.

3. Simulation results

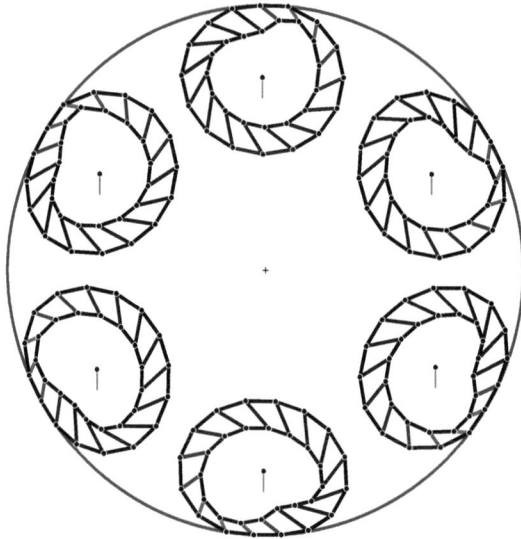

Figure 3.19.: The doube-chain model moving in a tube, demonstrating the wall and ceiling climbing capabilities of the model. Figure shows six positions during one rotation in the tube in counter-clockwise direction. The vector at the centre point is the gravity vector, showing the direction and magnitude of gravity. Shades of red colouring define the elasticity gradient.

3.4. Capabilities

The final test for the robot model is their performance and capabilities against the direction of gravity. If a robot model is able to move upwards an inclination it can also transport a load. One step further, if the adhesion is strong enough, the weight of the model low enough and the generated traction forces high enough, the model should be able to climb along walls and maybe also on overhanging inclinations or even on ceilings. For a simple simulation test, the model is put into a circular tube, which is defined by polar coordinates with the centre point of the tube as origin. To give the models enough room to move, the tube has a radius of 10 length units. Initially, the model is put in the tube at the bottom, the lowest point, where gravity is orthogonal to the surface (defined by polar angle $\theta = -\pi/2$).

The double chain model is extraordinarily performing in this test setup – It is able to move one full rotation in the tube without stopping, rolling back or dropping off [Figure 3.19] with the given set of standard parameters [Table 2.1]. Analysing the angular speed (the difference of polar angle θ over time) in relation to the current position of the

3.4. Capabilities

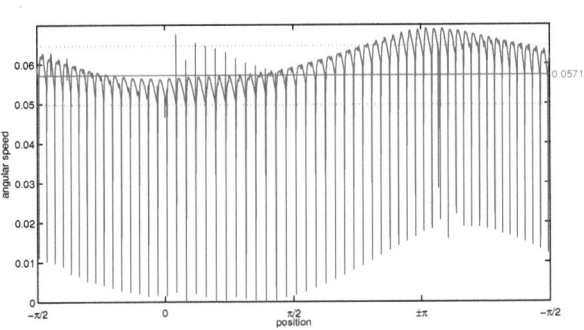

Figure 3.20.: The angular speed of the double-chain model inside the tube during on rotation in relation to the position of the circular tube, defined by angle θ. The grey line marks the mean angular speed of one rotation with the dotted line defining the interval of its standard deviation.

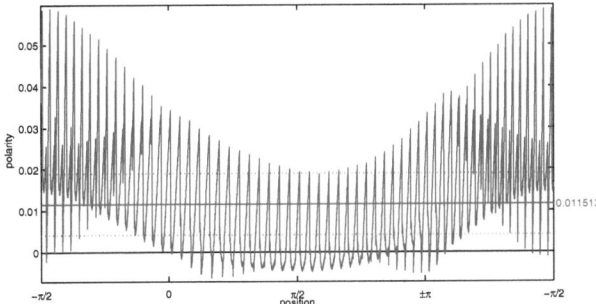

Figure 3.21.: The polarity of the double-chain model inside the tube in relation to the position of the circular tube, defined by angle θ. The grey line marks the mean polarity of one rotation with the dotted line defining the interval of its standard deviation.

model in the tube (defined by polar angle θ [Figure 3.20]) reveals the tube regions of acceleration and deceleration. Despite the fluctuations caused by detachment of vertices, the minimal angular speed is approximately reached at position $\theta = 0$, the position, where gravity is acting in exact opposite direction of movement (comparable with moving upwards a wall with 90 degree inclination). The model is accelerating faster after reaching the position at polar angle $\theta = \pi/2$, defining the highest point in the tube. Beyond this point the angle between the movement direction and the gravity vector becomes smaller than 90 degrees. Gravity is now pulling the model in direction of movement. Gravity and movement direction are parallel at position $\theta = \pm\pi$, which results in the maximum angular speed shortly after that position. This results in an asymmetric angular speed graph [Figure 3.20].

67

3. Simulation results

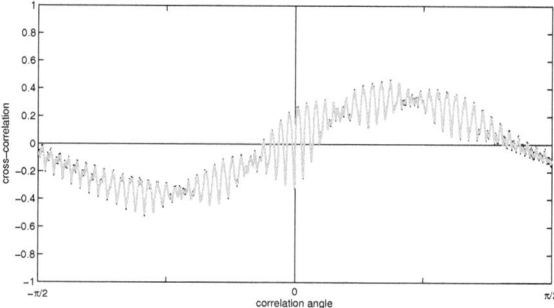

Figure 3.22.: The cross-correlation between the angular speed [Figure 3.20] and polarity [Figure 3.21] in relation to the angular position of the model. There is a positive correlation at $\sim \pi/3$. Black dotted lines define the confidence interval of correlation.

The direction of gravity will influence the polarity of the double chain model. This is shown in Figure 3.21 with the polarity at each position in the tube, defined by polar angle θ. It demonstrates, that the lowest polarity is reached shortly after the highest position in the tube after $\theta = \pi/2$, when the model is on its way down again, whereas at the bottom of the tube the polarity reaches its maximum. The direction of gravity is also enhancing polarity. If gravity is pushing the model to the surface, the polarity is higher than in the other case, where gravity is pulling the model away from the surface (which results in easier detachment of vertices). This results in an asymmetric angular polarity graph [Figure 3.21].

Figure 3.22 depicts the cross-correlation between both graphs (the angular speed and the angular polarity) in relation to the angular position. There is a positive correlation at approximately $\pi/3$, the polarity is following the the angular speed with an angle offset of approximately $\pi/3$. This corresponds also with a temporal delay.

The extended single-chain model can take certain inclinations but is not able to climb higher than the bottom right position of the double-chain model in Figure 3.19. Since that position has about 50 degree of inclination, the performance is already quite good for such a simple model.

In summary, the extended-single chain model and the double-chain model can both take inclinations, therefore both are also able to transport some load. The double-chain model is a formidable climber, taking every angle without problems. This analysis also revealed, that the direction of gravity can enhance polarity.

4. Concluding evaluation

According to the simulation results, each of the introduced simulation models of a bio-inspired locomotion device demonstrated its functionality and proved the possibility to adapt the biophysical model of cell migration for a bionic application. In this case the transferred and applied primary biological mechanism for this locomotion device is the observed transition of the cytoskeletal actin cortex, which is one of the driving mechanisms of the migration of cells. This transition creates a visco-elastic gradient and polarises the cell by defining a "sloppy" leading edge, where the cortex is more a sol, a solution-like viscous material, built by a rough flexible network of actin filaments and a "stiff" rear end, where the cortex is more a gel, a solid-like elastic material, consisting of rigid bundles of actin fibres, which is confirmed by simulation studies and experimental data [6, 37]. This creation of the gradient and the accompanying polarisation can be considered as a biological self-organising process. It is an autonomous self-contained system, a self-amplifying autokatalytic mechanism, autonomously driving and enhancing itself once started. By adaptation of this mechanism into the models, they inherit these self-organising and self-amplifying properties. That is advantageous for a technical application, because there is no need for an external control to start or maintain the locomotion of the device. It only requires an adhesion-dependent trigger for activation. Considering a technical adaptation this gradient is simplified as an elasticity gradient which is easier to technically implement, requiring only actuators that are able to change their elastic properties.

Simulations showed that an elasticity gradient is sufficient to propel the models and uncovered some distinctive observable differences between the introduced models. The simplest of the introduced models is the single-chain model. Its elasticity gradient is basically defined by the temporally change of the bending stiffness of the chain. It is able to move with a persistent locomotion speed, but is the worst performer of all models, so that its locomotion can be described as creeping with the lowest translocation speed and very susceptible to instability. Parameter scans also revealed that there is not much room to improve its performance by changing elasticity related parameters. Nevertheless, it is the simplest model and has the advantage for easy implementation of new features for testing purposes, but the technical realisability is complicated, because all the required mechanics have to be built into one chain, implemented into the segments and vertices.

The extended single-chain is an improvement of the single-chain model by adding elastic spikes extending from the outer side of the chain. The elasticity gradient is additionally defined by the temporal change of the elastic properties of these spikes. It also requires a lower bending stiffness. The extended single-chain comes second on the best performer list with its moderate performance of a crawling locomotion, with a ten times higher translocation speed compared to the single-chain model and improved stability and moderate climbing abilities. The main advantage of the extended single-chain model is the possible adaptation to higher roughness with its ability to compensate unevenness by using the spikes as feet and by stepping over small obstacles, which can be described as some type of "floating" over roughness.

The double-chain model is a further improvement of the previous model, consisting of two chains connected with elastic spokes. The elasticity gradient is primarily defined by the temporal change of the elastic properties of the spokes. It is the performance winner, demonstrating a rolling movement with the highest observed translocation speed and high stability and the used parameters even allow for further enhancements. The crawling capabilities are extraordinary, the double-chain model has the ability to move against any direction of gravity allowing to move along any (overhanging) inclination.

The different types of locomotion also show observable differences, especially between the crawling of the extended single-chain model and the rolling of the double-chain model. One difference is the distribution of forces along the attached vertices. The crawling locomotion has traction primary on the front part, where the applied horizontal force at the rear end is pointing in direction of movement, which is counterproductive for a net forward traction but improves forward locomotion by enhancing retraction after detachment of the last attached vertex. The traction of the rolling locomotion is applied at the rear end, but this is disadvantageous for retraction. This locomotion type is accompanied with a special shape deformation. The inner chain is polarised, while the outer chain is compensating the polarity of the inner chain, which explains the observed shear of the two chains. In contrast, the crawling locomotion is accompanied by a polarity in shape with a flattened front.

4.1. Constructability

The precondition for real construction of a prototype is fulfilled by the non-dimensionalisation and the simplification of the introduced simulation models. Although their intrinsic properties require certain smart materials, which meet the qualities postulated in this work (particularly the elastic properties). Especially special linear actuators are required, which change their elastic properties (or in other means their length) during activation by adhesion. An actuator wish list for the best material properties include the following features: small-scale, light-weight, high force generation, high displacement, high frequency operation, low hysteresis and low energy requirements. Unfortunately, such super-material or actuator is not available. But that is no reason to be worry. Not all of the listed properties are equally important in this case. A short compilation of available smart materials and actuators which summarises their working principle, qualities and disadvantages can be found at section A.2. Evaluating the available qualities and disadvantages in relation to the required properties of the simulation models yields the most important feature an actuator must have: the ability to generate forces with a high stroke or displacement. The least important features are the ability to operate with high frequency and low hysteresis. High frequency is unimportant because the time between activation and deactivation should be long enough. Low hysteresis is not that important, because the actuators do not have to work very precisely and possible creep gets compensated by the flexibility. In case of weight and scale it depends on the final size of the model.

For a small scale prototype, SMA actuators are offering the best properties for realisation and implementation [section A.2]. SMAs are already used as actuators in smaller scale robot prototypes [92] and field-tested in robotics. Their advantage for usage in a small scale robot is their light-weight properties and the ability for high strokes. Besides, they are easy to implement, requiring only a power source for operation. SMA springs are special springs made of SMA materials, they are available as tensile and compression springs with capabilites of high force generation and large strokes [124]. They can be connected with an ordinary spring (see figures in [124]). Both springs in series are acting as one single elastic element and actuator. A power source connected to the SMA spring is triggering its thermoelastic effect and both springs calibrate to a new stiffer equilibrium elasticity. The activation mechanism can also be easily implemented, requiring only a small pin, which is getting pushed during attachment and triggers the power source.

On a larger scale the electroactive polymer actuators [section A.2] might be an option for realisation.

An alternative and relatively new concept for actuation is using the effect of particle jamming. Jamming describes the physical process by which granular materials become more rigid by increasing its density. It is proposed as a new type of phase transition [17]. Increasing the density prevents the particles from exploring phase space, they become jammed, and the material behaves like a solid. This effect is reversible, the material is able to unjam by increasing temperature or applying external stress. The jamming phase transition relates to inverse density, stress and temperature [97].

This effect is used in new soft actuators, such as the Jamming Modulated Unimorph (JMU) [90] and in a new concept of jamming skin enabled locomotion [91]. Schematics of this concept is shown in Figure 4.1 and the corresponding robot prototype, which was built by the company iRobot, in Figure 4.2. This robot is comprised of many cellular compartments that enclose a fluid-filled cavity (or in the simplest case air-filled). The cellular compartments contain jamming material each of which can be jammed (therefore increasing rigidity) by applying a vacuum or unjammed (increasing flexibility) by releasing the vacuum. The central fluid-filled cavity is the only actuator, pumping a fluid or air into this cavity is the actuation mechanism. This robot is able to perform a simple rolling gait [91, 90]. This robot does look similar to the introduced double-chain model and the working principle should be easily adaptable for the bio-inspired propulsion concept, introduced in this work.

4. Concluding evaluation

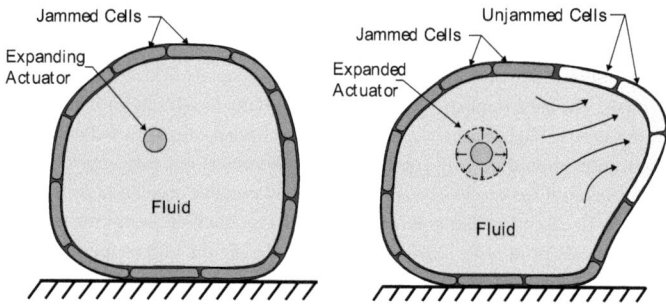

Figure 4.1.: Jamming skin enabled locomotion (JSEL) topology both unactuated (left) and actuated with a subset of the cells jammed (right). Figure taken from Steltz et al. [91].

Figure 4.2: The robot prototype as proof of concept for jamming skin enabled locomotion. A robot is comprised of 20 cellular compartments that enclose an air-filled cavity. The cellular compartments contain jamming material each of which can be jammed (increasing rigidity) by applying a vacuum or unjammed (incrfeasing flexibility) by releasing the vacuum. The central cavity is the only actuator, pumping air into this cavity is the actuation mechanism. Figure taken from Steltz et al. [91].

For realisation of the required adhesion mechanism: there was no particular adhesion model used in this work, except a simple adhesion by touching the surface. Adhesion has to withstand strong horizontal forces (as seen in the simulations), hence de-adhesion is only possible in vertical direction. This should be considered in the technical realisation of the adhesion mechanism.

4.2. Usability

After proving the functionality and the possible constructability, what is the usability of these models? Primarily, this work wants to demonstrate the technical feasibility of a bio-inspired or bionic locomotion device based on cell migration: the demonstration of a bio-inspired climbing robot independent of the disadvantages of legs as seen in many other bio-inspired climbing robots, such as geckobots [99] and stickybots [53, 86, 9, 98] (but still using adhesion for climbing) and combined with the flexible shape of soft robots. Additionally, the robot models are autonomous in their locomotion. There is no need for an external control and the movement is simply triggered by adhesion compared to the deformable soft robot prototype of Sugiyama and Hirai introduced at the beginning [92] (which is similar in shape and also using a type of crawling/rolling locomotion), which is needing applied voltage patterns controlling the SMA coils for movement, whereas this control pattern is simply created by the intrinsic gradient in the bio-inspired models.

Nonetheless the possible field of application for the introduced bio-inspired locomotion device are environments, where the classic wheeled and legged locomotion is limited, this includes area with uneven terrain, obstacles and uncertain or changing conditions or environments with low gravity or even no gravity. This comprises conditions where the flexible shape and the adhesion-based locomotion is most advantageous. The double-chain model demonstrated its ability to move inside a tube, defying the influence of gravity. A similar environment is the inside of pipelines. Therefore one possible application is a locomotion device for a cleaning robot for pipelines. Another idea is a bionic concept: The introduced locomotion device is inspired by cell migration. Keratinocytes, tissue cells of the skins, are known for wound healing, they migrate to damaged sites of the skin and proliferate for reepithelisation [85]. As a new bionic technical concept: the introduced robot model can serve as an emergency repair robot on spacecrafts. This robot can move along the outer hull (the "skin") of a spacecraft to a leakage or damaged part of the hull (the "wound") and provisorily seal the leakage with its body until it can be repaired (the "healing"). The locomotion is dependent on adhesion and therefore independent of gravity and perfectly suited to space environment.

4.3. Reverse bionics

Reverse bionics is the process to learn something about the natural biological model from bionic technology. In this case, the simulations have the advantage that every involved parameter influencing the system is known, which is not given for a biological system. So what insights about the biological model can be gained by this work through the process of reverse bionics?

Simulations with rough surfaces revealed that the extended single-chain model with its filopodium-like spikes has advantages on rougher surfaces, whereas the double-chain model performs better with less roughness. The first hypothesis: the different structures of the leading edge, such as the broad lamellipodium, the tubular filopodia or the pseudopodia projections (as seen on different migrating cell types), are adaptations to the natural environment where these cells normally migrate in vivo. The broad lamellipodium seems to be made for a rather flat and even surface, whereas filopodia seem to be made for rough or porous surfaces. Pseudopodia are used by singulary living and migrating cells, maybe a universal adaptation to uncertain conditions of the surface?

Another observation during simulation is the fast rolling locomotion of the double-chain model (with less adhesion points), which lead to the second hypothesis: fast moving cells utilize some kind of rolling movement with a lamellipodial structure (like rolling out a carpet on the floor). Fish keratocytes are very fast moving cells with a lamellipodium and weak adhesion [4], and experimental data suggest that their cell body is rolling during migration [8, 93].

The force analysis revealed that the extended single-chain model is crawling with traction force mainly generated at the front of the cell. Third hypothesis: crawling cells with filopodial structures have in sum a higher traction force at the front. Comparing this to experimental data, the crawling fibroblasts with their filopodial structures at the leading edge seem to have indeed larger traction forces at the front [27, 70, 71].

4.4. Outlook

As seen during simulations, the extended single-chain model has advantages on rough surfaces due to its extending spikes, while the double-chain model is the overall winner in case of performance, especially with its climbing capabilities. The next step is the combination of both models, resulting in an extended double-chain model: The double-chain model will serve as body and motorisation, but is extended by spikes as seen in the extended single-chain model. This model should unite the advantages of both models, the adaptive behaviour to rough surfaces and the extraordinary climbing ability.

Another future step is the realisation of the models with a prototype. The simulations proved the basic functionality. Besides, adequate concepts and materials for construction are available, especially the jamming skin enabled locomotion looks promising for a technical realisation of the models, since this physical jamming effect is similar to the observable cytoskeletal phase transition during cell migration.

Nonetheless this work introduced an interesting concept of a bio-inspired and bionic robot locomotion device that is able to crawl or roll for locomotion. Further improvement and fine-tuning may lead to a new generation of soft robots.

A. Appendix

A.1. Mathematical derivation of friction models

Drag dependent friction

A moving segment is causing drag, which counteracts the velocities of the two vertices of this segment. For a single segment, this drag can be described with an orthogonal and a parallel component. Considering the orthogonal direction, the drag is assumed to increase over the length of the segment the further away this point is from the vertex of interest. Taking the vertex Z_i as vertex of interest, it has two neighbouring segments in the single chain model, one neighbour is the segment $i\text{-}1$ with length L_{i-1} and with vertex Z_{i-1} at the other segment's end. Discretising the length L_{i-1} into s parts, the orthogonal component depending on the orthogonal velocity component $v^\perp$ can be written as:

$$f^\perp(L_{i-1}) = \int_0^{L_{i-1}} \frac{1}{2}\frac{s}{L_{i-1}}[v_{i-1}^\perp(1-\frac{s}{L_{i-1}}) + v_i^\perp(\frac{s}{L_{i-1}})]\, ds$$

Solving the integral is resulting in:

$$f^\perp(L_{i-1}) = \frac{1}{6}L_{i-1}\, v_i^\perp + \frac{1}{12}L_{i-1}\, v_{i-1}^\perp$$

and for the other segment accordingly:

$$f^\perp(L_i) = \frac{1}{6}L_i\, v_i^\perp + \frac{1}{12}L_i\, v_{i+1}^\perp$$

Thus the normal friction force for both segments can be written as:

$$F_{f,L_{i-1}}^\perp = -\eta_n \vec{n}_{i-1}(\frac{1}{6}L_{i-1}\, v_i^\perp + \frac{1}{12}L_{i-1}\, v_{i-1}^\perp)$$

$$F_{f,L_i}^\perp = -\eta_n \vec{n}_i(\frac{1}{6}L_i\, v_i^\perp + \frac{1}{12}L_i\, v_{i+1}^\perp)$$

with friction coefficient in normal direction η_n and normal unit vector $\vec{n}$.

Combining the two formula, the friction in normal direction at vertex Z_i yields:

$$F_{f,i}^\perp = -\frac{\eta_n}{6}(L_{i-1}\vec{n}_{i-1}\, v_i^\perp + L_i\vec{n}_i\, v_i^\perp) - \frac{\eta_n}{12}(L_{i-1}\vec{n}_{i-1}\, v_{i-1}^\perp + L_i\vec{n}_i\, v_{i+1}^\perp)$$

A. Appendix

The parallel component is working similar as the one in orthogonal direction but with the difference that drag/friction is assumed to be uniform in parallel direction and therefore independent of the segment's length. Thus, the formula for the friction in tangential direction at vertex Z_i is:

$$F_{f,i}^{\parallel} = -\frac{\eta_u}{2}(\vec{u}_{i-1}\ v_i^{\parallel} + \vec{u}_i\ v_i^{\parallel}) - \frac{\eta_u}{2}(\vec{u}_{i-1}\ v_{i-1}^{\parallel} + \vec{u}_i\ v_{i+1}^{\parallel})$$

Additionally, the segments' elasticity causes additional friction in parallel direction, expressed in a formula at vertex Z_i:

$$F_{fe,i}^{\parallel} = -\eta_e(\vec{u}_{i-1}\ v_i^{\parallel} + \vec{u}_i\ v_i^{\parallel}) + \eta_e(\vec{u}_{i-1}\ v_{i-1}^{\parallel} + \vec{u}_i\ v_{i+1}^{\parallel})$$

Replacing $v^{\perp}$ by $\vec{v}\cdot\vec{n}$ and $v^{\parallel}$ by $\vec{v}\cdot\vec{u}$ and uniting the three formulas into one is giving the full drag friction at vertex Z_i:

$$\vec{F}_{f,i} = -\frac{\eta_n}{6}[L_{i-1}\vec{n}_{i-1}(\vec{n}_{i-1}\ \vec{v}_i) + L_i\vec{n}_i(\vec{n}_i\ \vec{v}_i)] - \frac{\eta_n}{12}[L_{i-1}\vec{n}_{i-1}(\vec{n}_{i-1}\ \vec{v}_{i-1}) + L_i\vec{n}_i(\vec{n}_i\ \vec{v}_{i+1})]$$
$$-\frac{\eta_u}{2}[\vec{u}_{i-1}(\vec{u}_{i-1}\ \vec{v}_i) + \vec{u}_i(\vec{u}_i\ \vec{v}_i)] - \frac{\eta_u}{2}[\vec{u}_{i-1}(\vec{u}_{i-1}\ \vec{v}_{i-1}) + \vec{u}_i(\vec{u}_i\ \vec{v}_{i+1})]$$
$$-\eta_e[\vec{u}_{i-1}(\vec{u}_{i-1}\ \vec{v}_i) + \vec{u}_i(\vec{u}_i\ \vec{v}_i)] + \eta_e[\vec{u}_{i-1}(\vec{u}_{i-1}\ \vec{v}_{i-1}) + \vec{u}_i(\vec{u}_i\ \vec{v}_{i+1})]$$

and collecting the summands results in the final equation:

$$\vec{F}_{f,i} = -[\frac{\eta_n}{12}[L_{i-1}(\vec{n}_{i-1}\cdot\vec{n}_{i-1}^T)] + (\frac{\eta_u}{2} + \eta_e)[(\vec{u}_{i-1}\cdot\vec{u}_{i-1}^T)]]\ \vec{v}_{i-1}$$
$$-[\frac{\eta_n}{6}[L_{i-1}(\vec{n}_{i-1}\cdot\vec{n}_{i-1}^T) + L_i(\vec{n}_i\cdot\vec{n}_i^T)] + (\frac{\eta_u}{2} - \eta_e)[(\vec{u}_{i-1}\cdot\vec{u}_{i-1}^T) + (\vec{u}_i\cdot\vec{u}_i^T)]]\ \vec{v}_i$$
$$-[\frac{\eta_n}{12}[L_i(\vec{n}_i\cdot\vec{n}_i^T)] + (\frac{\eta_u}{2} + \eta_e)[(\vec{u}_i\cdot\vec{u}_i^T)]]\ \vec{v}_{i+1}$$

Moment dependent friction

The angular friction moment at each vertex' torsion spring depends on the temporal change of the angle φ between two segments. The sinus of angle φ can be defined as the product of vectors $\vec{n}$ (the inner normal unit vector of segments) and $\vec{u}$ (directional unit vector of segments):

$$-\vec{n}_{i-1}\cdot\vec{u}_i = \sin(\varphi)$$

which can be transformed to

$$-\frac{\vec{Z}_i^{\perp} - \vec{Z}_{i-1}^{\perp}}{L_{i-1}} \cdot \frac{\vec{Z}_{i+1} - \vec{Z}_i}{L_i} = \sin(\varphi) =: s\varphi$$

A.1. Mathematical derivation of friction models

thus
$$(\vec{Z}_{i-1}^{\perp} - \vec{Z}_i^{\perp}) \cdot (\vec{Z}_{i+1} - \vec{Z}_i) = L_{i-1} \cdot L_i \cdot s\varphi$$

with position vector $\vec{Z}$ of the vertices and the segments' length L.

The time derivative of the previous equation provides a relation to the velocities $\vec{v} = \dot{\vec{Z}}$ of the free vertices:

$$(\vec{v}_{i-1}^{\perp} - \vec{v}_i^{\perp})(\vec{Z}_{i+1} - \vec{Z}_i) + (\vec{v}_{i+1} - \vec{v}_i)(\vec{Z}_{i-1}^{\perp} - \vec{Z}_i^{\perp}) = \left(\frac{\dot{L}_{i-1}}{L_{i-1}} + \frac{\dot{L}_i}{L_i}\right)(\vec{Z}_{i-1}^{\perp} - \vec{Z}_i^{\perp})(\vec{Z}_{i+1} - \vec{Z}_i) + L_i \cdot L_{i-1} \cdot s\dot{\varphi}$$

(with velocity vector $\vec{v}$).

Dividing this equation by $(L_{i-1} \cdot L_i)$ results in:

$$\frac{1}{L_{i-1}}(\vec{v}_{i-1}^{\perp} - \vec{v}_i^{\perp}) \cdot \vec{u}_i + \frac{1}{L_i}(\vec{v}_i - \vec{v}_{i+1}) \cdot \vec{n}_{i-1} + \left(\frac{\dot{L}_{i-1}}{L_{i-1}} + \frac{\dot{L}_i}{L_i}\right)(\vec{n}_{i-1} \cdot \vec{u}_i) = s\dot{\varphi}$$

and, because $(\vec{v}_{i-1}^{\perp} - \vec{v}_i^{\perp}) \cdot \vec{u}_i = (\vec{v}_i - \vec{v}_{i-1}) \cdot \vec{n}_i$,

$$\frac{1}{L_{i-1}}(\vec{v}_i - \vec{v}_{i-1}) \cdot \vec{n}_i + \frac{1}{L_i}(\vec{v}_i - \vec{v}_{i+1}) \cdot \vec{n}_{i-1} + \left(\frac{\dot{L}_{i-1}}{L_{i-1}} + \frac{\dot{L}_i}{L_i}\right)(\vec{n}_{i-1} \cdot \vec{u}_i) = s\dot{\varphi}$$

This equation is further transformed by $\dot{L}_{i-1} = (\vec{v}_i - \vec{v}_{i-1}) \cdot \vec{u}_{i-1}$ and $\dot{L}_i = (\vec{v}_{i+1} - \vec{v}_i) \cdot \vec{u}_i$:

$$\frac{1}{L_{i-1}}(\vec{v}_i - \vec{v}_{i-1}) \cdot \vec{n}_i + \frac{1}{L_i}(\vec{v}_i - \vec{v}_{i+1}) \cdot \vec{n}_{i-1}$$
$$+\left[\frac{1}{L_{i-1}}(\vec{v}_i - \vec{v}_{i-1}) \cdot \vec{u}_{i-1} + \frac{1}{L_i}(\vec{v}_{i+1} - \vec{v}_i) \cdot \vec{u}_i\right] \cdot (\vec{n}_{i-1} \cdot \vec{u}_i) = s\dot{\varphi}$$

obtaining

$$\frac{1}{L_{i-1}}(\vec{v}_i - \vec{v}_{i-1}) \cdot \vec{n}_i + \frac{1}{L_i}(\vec{v}_i - \vec{v}_{i+1}) \cdot \vec{n}_{i-1}$$
$$+\left[\frac{1}{L_{i-1}}(\vec{v}_i - \vec{v}_{i-1}) \cdot \vec{u}_{i-1} - \frac{1}{L_i}(\vec{v}_i - \vec{v}_{i+1}) \cdot \vec{u}_i\right] \cdot (\vec{n}_{i-1} \cdot \vec{u}_i) = s\dot{\varphi}$$

finally yielding

$$\frac{1}{L_{i-1}}[\vec{n}_i + \vec{u}_{i-1} \cdot (\vec{n}_{i-1} \cdot \vec{u}_i)] \cdot (\vec{v}_i - \vec{v}_{i-1})$$
$$+ \frac{1}{L_i}[\vec{n}_{i-1} - \vec{u}_i \cdot (\vec{n}_{i-1} \cdot \vec{u}_i)] \cdot (\vec{v}_i - \vec{v}_{i+1}) = s\dot{\varphi}$$

The friction of the moment at vertex Z_i is assumed to be

$$M = -\eta \cdot \dot{\varphi} = -\tilde{\eta} \cdot s\dot{\varphi}$$

A. Appendix

with
$$\tilde{\eta} = \eta \cdot \frac{1}{\cos(\varphi)}$$

Thus, the final friction term of the angular moment at vertex Z_i is:
$$M_{f,i} = -\tilde{\eta}_i \left[\vec{m}_i^+ \left(\vec{v}_i - \vec{v}_{i-1} \right) + \vec{m}_i^- \left(\vec{v}_i - \vec{v}_{i+1} \right) \right]$$

with $\vec{m}_i^+ = \frac{1}{L_{i-1}}[\vec{n}_i + \vec{u}_{i-1}(\vec{n}_{i-1} \cdot \vec{u}_i)]$ and $\vec{m}_i^- = \frac{1}{L_i}[\vec{n}_{i-1} - \vec{u}_i(\vec{n}_{i-1} \cdot \vec{u}_i)]$.

The total friction force at vertex Z_i is affected by the sum of moments of each neighbouring segments, therefore the previous equation is plugged in the formula for the sum of moments at each vertex [see Equation 2.8].

$$\vec{F}_{f,i} = \frac{M_{f,i}}{L_i}\vec{n}_i + \frac{M_{f,i}}{L_{i-1}}\vec{n}_{i-1} - \frac{M_{f,i+1}}{L_i}\vec{n}_i - \frac{M_{f,i-1}}{L_{i-1}}\vec{n}_{i-1}$$

or

$$\vec{F}_{f,i} = -\tilde{\eta}_i \frac{1}{L_i} \left[\vec{m}_i^+ \left(\vec{v}_i - \vec{v}_{i-1} \right) + \vec{m}_i^- \left(\vec{v}_i - \vec{v}_{i+1} \right) \right] \vec{n}_i$$
$$- \tilde{\eta}_i \frac{1}{L_{i-1}} \left[\vec{m}_i^+ \left(\vec{v}_i - \vec{v}_{i-1} \right) + \vec{m}_i^- \left(\vec{v}_i - \vec{v}_{i+1} \right) \right] \vec{n}_{i-1}$$
$$+ \tilde{\eta}_{i+1} \frac{1}{L_i} \left[\vec{m}_{i+1}^+ \left(\vec{v}_{i+1} - \vec{v}_i \right) + \vec{m}_{i+1}^- \left(\vec{v}_{i+1} - \vec{v}_{i+2} \right) \right] \vec{n}_i$$
$$+ \tilde{\eta}_{i-1} \frac{1}{L_{i-1}} \left[\vec{m}_{i-1}^+ \left(\vec{v}_{i-1} - \vec{v}_{i-2} \right) + \vec{m}_{i-1}^- \left(\vec{v}_{i-1} - \vec{v}_i \right) \right] \vec{n}_{i-1}$$

collecting the summands results to the final equation:

$$\vec{F}_{f,i} = -[\tilde{\eta}_{i-1} \frac{1}{L_{i-1}} \vec{n}_{i-1} \cdot \vec{m}_{i-1}^+] \vec{v}_{i-2}$$
$$+ [\tilde{\eta}_i (\frac{1}{L_i} \vec{n}_i + \frac{1}{L_{i-1}} \vec{n}_{i-1}) \cdot \vec{m}_i^+ + \tilde{\eta}_{i-1} \frac{1}{L_{i-1}} \vec{n}_{i-1} \cdot (\vec{m}_{i-1}^+ + \vec{m}_{i-1}^-)] \vec{v}_{i-1}$$
$$- [\tilde{\eta}_i (\frac{1}{L_i} \vec{n}_i + \frac{1}{L_{i-1}} \vec{n}_{i-1}) \cdot (\vec{m}_i^+ + \vec{m}_i^-) + \tilde{\eta}_{i+1} \frac{1}{L_i} \vec{n}_i \cdot \vec{m}_{i+1}^+ + \tilde{\eta}_{i-1} \frac{1}{L_{i-1}} \vec{n}_{i-1} \cdot \vec{m}_{i-1}^-] \vec{v}_i$$
$$+ [\tilde{\eta}_i (\frac{1}{L_i} \vec{n}_i + \frac{1}{L_{i-1}} \vec{n}_{i-1}) \cdot \vec{m}_i^- + \tilde{\eta}_{i+1} \frac{1}{L_i} \vec{n}_i \cdot (\vec{m}_{i+1}^+ + \vec{m}_{i+1}^-)] \vec{v}_{i+1}$$
$$- [\tilde{\eta}_{i+1} \frac{1}{L_i} \vec{n}_i \cdot \vec{m}_{i+1}^-] \vec{v}_{i+2}$$

with $\vec{m}_i^+ = \frac{1}{L_{i-1}}[\vec{n}_i + \vec{u}_{i-1}(\vec{n}_{i-1} \cdot \vec{u}_i)]$ and $\vec{m}_i^- = \frac{1}{L_i}[\vec{n}_{i-1} - \vec{u}_i(\vec{n}_{i-1} \cdot \vec{u}_i)]$.

A.2. Smart material actuators

In engineering and material sciences there are a few options of smart material actuators available for usage, each with certain advantages and disadvantages.

Piezoelectric actuators

Piezoelectric actuators are small-scale and relatively stiff, load-bearing and stackable actuators. It is a well understood and commercialised technology. They are also usable for sensorics but the usage for actuation is more common. Those actuators are made of piezoceramics – a popular material is lead zirconate titanate (PZT). Their operating principle allow dynamic strains (capable of fine positioning even on a micro-scale atomic level) and oscillatory applications. Therefore, typical practical applications include relays, microphones and loudspeakers, inkjet printers, strain gauges and especially in atomic force microscopes and scanning tunnelling microscopes to fine-tune the position of the microscope's head.

Principle   Piezoceramics respond to electric fields and experience mechanical deformations when exposed to them due to the internal crystal structure of a piezoceramic. It has no center of symmetry, so ions in this anisotropic crystal lattice can be displaced by an outside electric field, resulting in a polarised material. This diplacement of charges is linked to a displacement in the crystal lattice. These atomic spatial displacements sum up to a deformation of the whole material.

Qualities   Piezoelectric actuators typically exert a displacement from 0.1 to 0.2% strain with a good linearity, possible in gigahertz frequency range. They are electrically driven, allowing them directly integrated and controlled by the electronics of the technical device. The material is moderately priced in comparison to other actuator technologies. Their physical properties include a low thermal coefficient, a density around 7.5 to 7.8 $\times 10^3$ kg m^{-3} and a maximum operating temperature near 300°C [39].

Disadvantages   Their main disadvantage is their high voltage requirements, typically in a range from 1 to 2 kV, which scales with the size of the actuator – as the size increases, so does the voltage. This limits their optimal application to small-scale devices. Other disadvantages are their high hysteresis and creep around 15 to 20%.

Single-crystal piezoceramics are a new development to improve current piezoceramics and counterbalance some of their disadvantages, offering a lower hysteresis (around 10 to 15%) and 5 to 10 times greater strain, though they are costlier to manufacture [43].

Magneto- and electrostrictors

Magnetostrictors are large-scale, high-force and high-stiffness actuators. They elongate in direction of an applied uniform longitudinal magnetic field. Electrostrictors function similar to magnetostrictors but are controlled by an applied electric field, thus they are used in a similar fashion like piezocelectric actuators with the main difference that electrostrictive actuators experience deformation in direction and orthogonal to the direction of the electric field, while piezoceramics are bi-directional (physically the piezoelectric effect is related to electrostriction). The highest known magnetostriction is exhibited by Terfenol-D, a material composed of terbium, iron, and the expensive rare-earth dysprosium, while electrostrictive behaviour is observable on all dielectric materials.

Principle Magentostriction is an effect due to intrinsic magnetic domains within the material. These domains rotate to align with an applied magnetic field, which distorts the crystal structure. In detail the formation of the more or less random aligned magnetic domains to an ordered alignment along the magnetic field allows a proportional, fast and repeatable expansion of the material. The displacement per unit magnetic field increases with dimension, therefore magnetostrictive materials allow for large-scale and heavy-duty actuators.

Electrostriction is caused by polar domains within the dielectric material. By applying an electrical field the opposite sides of the domains become differently charged and attracting each other, resulting in a reduction of material thickness in the direction of the applied electric field and increased thickness in the orthogonal direction of the field (characterised by Poisson's ratio). The strain is proportional to the square of the polarisation. Since electrostrictive effects are present in nearly all materials, only those with large effects (> 0.7 nm per V) are useful as actuators [32].

Qualities Terfenol-D as best example for magnetostrictive materials is typically able to exert strains of 0.1 to 0.6% with operating frequencies from 0 to 30 kHz with a good linearity and a moderate hysteresis around 2%. The material has a density of about 9 $\times 10^3$ kg m^{-3} and has a maximum operating temperature near 400°C [39].

The most common electrostrictors are ceramics, which can provide a strain of 0.1 to 0.2% and operate from 20 to 100 kHz with an incredibly low hysteresis below 1%. Additionally, they have a low thermal coefficient and a density near 7.8 $\times 10^3$ kg m^{-3} and a maximum operating temperature near 300°C [32].

Disadvantages Magnetostrictive actuators require an applied controlled magnetic field – to create and maintain such a field requires more power than piezoelectric actuators. Additionally, if compressive load is applied to magnetostrictive materials, they tend to further interact with the device, which makes it more difficult to account this uncertain behaviour in planning and constructing the application. Finally, using Terfenol-D as magnetostrictive material may be not a pricey option because it requires a rather expensive rare-earth.

The main disadvantage of electrostrictors is their inherent nonlinearity. Their elongation follows a square law function of applied electric field. For compensation voltage biasing may be used to get regions of nominal linearity [32].

Shape-memory alloy (SMA) actuators

Shape-memory alloys (SMAs) are smart materials which are usable as thermal low-stiffness and high-displacement actuators. They are thermally activated and therefore their response time is more or less cooling dependent. The most common SMA material is Nitinol, which is typically fabricated as a wire for actuator use.

Principle Deformation of SMAs are based on a change of their intrinsic thermally dependent crystal lattice structure. Deformations of the crystal lattice during the martensic low-temperature phase revert back by 'heating' above a specific transformation temperature. Then the SMA will change its crystal structure to its austenic hig-temperature phase, 'remembering' its 'memorised' original shape. These phase transformation can not only be thermally induced but also by applying a current.

Qualities Forces and displacement is only limited by overall power. Theoretically, SMA actuators can provide infinitely high displacements or high forces (with a trade-off in near-zero force or near-zero displacement). Therefore, SMA materials can offer higher strains than any other smart actuator. Additionally, they have a good linearity and they are relatively simple to use – only the material and a current source is needed to operate them. Nitinol as popular SMA material is low-cost, has a density around 7×10^3 kg m^{-3} and a maximum operating temperature near 300°C [43]. Wires can be fabricated with around 50 micrometer in diameter. They are often used as micro-scale and macro-scale actuators in robotics. SMA materials can be bonded to other materials, producing bi-material cantilevers and actuators akin to many existing thermal actuators [108]. Another special usage of SMA materials is existing in the form of SMA springs. These special springs are made of shape-memory alloy and provide different elastic properties in their low- and high-temperature phase [124].

Disadvantages Besides high power requirements, the heating and cooling make them rather slow actuators, operating in a frequency between 0.5 and 5 Hz. Additionally, they have high hysteresis.

To encounter the low-frequency operation of SMA materials a newer development of smart materials are ferro-magnetic shape-memory alloys (FSMA), which are functioning

A. Appendix

similar to SMA but are magnetically activated and therefore operate faster than SMA actuators because no cooling is required. Though the trade-off is that additional structures are required to provide the magnetic field whereas SMA materials require only a current source.

Electroactive polymer (EAP) actuators

Wilhelm Conrad Röntgen (27th March 1845 – 10th February 1923) was one of the first, who discovered that certain types of polymers can change shape in response to electrical stimulation [84]. Electroactive polymers (EAPs) are smart material polymers that perform a change in size or shape when stimulated by an electric field, this effect is related to electrostriction mentioned earlier. They can exhibit high strains up to 380% with low energy requirements. In robotics they are used as artificial muscles [12]. EAPs can be divided into two groups: **Dielectric EAPs** (or Electronic EAPs) – comprising Dielectric Elastomer EAP, Electrostrictive Graft Elastomers, Electrostrictive Paper, Electro-Viscoelastic Elastomers, Ferroelectric Polymers and Liquid Crystal Elastomers (LCE); and **Ionic EAPs** – comprising Carbon Nanotubes (CNT), Conductive Polymers (CP), Electrorheological Fluids (ERF), Ionic Polymer Gels (IPG) and Ionic Polymer Metallic Composite (IPMC). The displacement of both types of EAPs can be geometrically designed to bend, stretch or contract.

Principle Dielectric EAPs are squeezed by electrostatic forces between two electrodes. Fundamentally, they are capacitors. When a voltage is applied, they change their capacitance and they compress in thickness and expand in area due to the electric field. Though this type of EAP typically requires a large actuation voltage to produce high electric fields, it consumes only very low electrical power. Therefore it has a high mechanical energy density. Additionally, it is operable in air. Such actuators are able to hold the induced displacement under activation and require no power to keep the actuator at a given position.

Ionic EAPs are driven by diffusion of ions – actuation is caused by the displacement of ions inside the polymer, therefore ionic EAPs need to be embedded in an electrolyte. As low as 1 – 2 Volts are needed for actuation, but the necessary ionic flow requires high electrical power and in contrast to electronic EAPs energy is needed to keep the actuator at a given position.

Qualities EAP materials are superior to shape memory alloys (SMA) in higher response speed, lower density, and greater resilience [12].

Dielectric EAPs exhibit high mechanical energy density, induce relatively large actuation forces, operate in room conditions, have a high response speed and can hold strain under activation [12].

Ionic EAPs have a natural bi-directional actuation dependent of voltage polarity, require only low voltage and some ionic polymers have a unique ability of bi-stability [12].

Disadavantages Dielectric EAPs are independent of the polarity of voltage, due to the related electrostriction effect they are mostly monopolar actuators. Besides, they require high voltages (~100 MV m^{-1}), though recent development with Ferroelectric EAPs requires only a fraction of the electric field [12].

Ionic EAPs require an electrolyte and their maintaining of wetness, because electrolysis occurs in aqueous environments. Thus, they need to be encapsulated with a protective layer in open air conditions. Additionally, they have a low electromechanical coupling efficiency. Except for CPs and NTs, they cannot hold strain without additional energy [12]. They have a slow response (fraction of a second). Bending Ionic EAPs induce only a low actuation force. Besides, it is difficult to manufacture a consistent materials (especially IPMC and with exception of CPs) [12].

B. Supplementary material

Additional supplementary material, such as animations of the locomotion process, the figures used in this work and the MATLAB® programme code of the models and simulations are available on CD-ROM.

The content is also available at

`http://bionic.chaos-engine.net`

References

[1] Abercrombie M (1980). The Croonian Lecture, 1978: The Crawling Movement of Metazoan Cells. *Proc Roy Soc Lond B.* 207:129–147.

[2] Abercrombie M, Heaysman JE, Pegrum SM (1970). The locomotion of fibroblasts in culture I. Movements of the leading edge. *Exp Cell Res.* 59(3):393–8.

[3] Abercrombie M, Heaysman JE, Pegrum SM (1970). The locomotion of fibroblasts in culture III. Movements of particles on the dorsal surface of the leading lamella. *Exp Cell Res.* 62(2):389-98.

[4] Alberts B, Johnson A, Lewis J, Raff M, Roberts K, Walter P (2002). *Molecular Biology of the Cell*, 4th Edition. Garland Science.

[5] Allen RD, Allen NS (1978). Cytoplasmic Streaming in Amoeboid Movement. *Annual Review of Biophysics and Bioengineering* 7:469–495.

[6] Alt W, Bock M, Möhl C (2010). In *Cell Mechanics: From Single Scale-Based Models to Multiscale Modelling*, Chauvière A, Preziosi L, Verdier C (editors) pp 86-125, Chapman & Hall.

[7] Ananthakrishnan R, Ehrlicher A (2007). The Forces Behind Cell Movement. *Int J Biol Sci* 3(5):303–317.

[8] Anderson KI, Wang YL, Small JV (1996). Coordination of protrusion and translocation of the keratocyte involves rolling of the cell body. *J. Cell Biol.* 134(5):1209–18.

[9] Asbeck A, Dastoor S, Parness A, Fullerton L, Esparza N, Soto D, Heyneman B, Cutkosky M (2009). Climbing rough vertical surfaces with hierarchical directional adhesion. *IEEE International Conference on Robotics and Automation*, pp. 2675–2680, 2009.

[10] Autumn K et al. (2002). Evidence for van der Waals adhesion in gecko setae. *PNAS* 99 (19): 12252–12256.

[11] Bandura J (2008). *Simulation eines mechanischen Roboter-Modells zur Zellmigration*. Diploma thesis – Theoretical Biology – University of Bonn, Germany.

[12] Bar-Cohen Y (2004). *Electroactive Polymer (EAP) Actuators as Artificial Muscles - Reality, Potential and Challenges*, 2nd Edition. SPIE Press, Vol. PM136.

[13] Bardy S, Ng S, Jarrell K (2003). Prokaryotic motility structures. *Microbiology* 149(2):295–304.

[14] Barthlott W, Neinhuis C (2001). The lotus-effect: nature's model for self cleaning surfaces. *International Textile Bulletin* 1: 8–12.

[15] Baumgartner W, Saxe F, Weth A et al. (2007). The Sandfish's Skin: Morphology, Chemistry and Reconstruction. *Journal of Bionic Engineering* 4(1):1–9.

[16] Bereiter-Hahn J, Lürs H (1998). Subcellular tension fields and mechanical resistance of the lamella front related to the direction of locomotion. *Cell Biochem. Biophys.* 29:243-262.

[17] Biroli G (2007). Jamming: A new kind of phase transition? *Nature Physics* 3(4):222-223.

[18] Block SM, Asbury CL, Shaevitz JW et al. (2003). Probing the kinesin reaction cycle with a 2D optical force clamp. *Proc Natl Acad Sci* 100:2351-2356.

[19] Brown ME, Bridgman PC (2003). Retrograde flow rate is increased in growth cones from myosin IIB knockout mice. *J Cell Sci* 116(6):1087-94.

[20] Brunner CA, Ehrlicher A, Kohlstrunk B et al. (2006). Cell migration through small gaps. *Eur Biophys J* 35(8):713-9.

[21] Cai Y, Biais N, Giannone G et al. (2006). Nonmuscle myosin IIA-dependent force inhibits cell spreading and drives F-actin flow. *Biophys J* 91(10):3907-20.

[22] Carlsson AE (2003). Growth velocities of branched actin networks. *Biophys J* 84(5):2907-18.

[23] Carlsson AE (2001). Growth of branched actin networks against obstacles. *Biophys J* 81(4):1907-23.

[24] Chiu H, Rubenstein M, Shen W-M (2008). 'Deformable Wheel' – A Self-Recovering Modular Rolling Track. Proc. Intl. Symposium on Distributed Robotic Systems, Tsukuba, Japan, November 2008.

[25] Crespi A, Ijspeert AJ (2006). AmphiBot II: an amphibious snake robot that crawls and swims using a central pattern generator. *Proceedings of the 9th International Conference on Climbing and Walking Robots (CLAWAR 2006)*, pp. 19-27.

[26] Degarmo EP, Black JT, Kohser RA (2003). *Materials and Processes in Manufacturing* 9th Edition, Wiley, p. 223.

[27] Dembo M, Wang YL (1999). Stresses at the cell-to-substrate interface during locomotion of fibroblasts. *Biophys J.* 76(4):2307-16.

[28] Den Outer A, Kaashoek JF, Hack HRGK (1995). Difficulties of using continuous fractal theory for discontinuity surfaces. *International Journal of Rock Mechanics and Mining Science & Geomechanics* Abstracts 32 (1):3-9.

[29] Diefenbach TJ, Latham VM, Yimlamai D et al. (2002). Myosin 1c and myosin IIB serve opposing roles in lamellipodial dynamics of the neuronal growth cone. *J Cell Biol* 158(7):1207-17.

[30] Dogterom M, Kerssemakers JW, Romet-Lemonne G et al. (2005). Force generation by dynamic microtubules. *Curr Opin Cell Biol.* 17(1):67-74.

[31] Dogterom M, Yurke B (1997). Measurement of the force-velocity relation for growing microtubules. *Science* 278(5339):856-60.

[32] Dorey AP, Moore JH (1995). *Advances in Actuators*, IOP Publishing.

[33] Dusenbery DB (2009). *Living at Micro Scale*. Harvard University Press, Cambridge, Mass.

[34] Euteneuer U, Schliwa M (1986). The Function of Microtubules in Directional Cell Movement. *Annals of the New York Academy of Sciences* 466(1):867-886.

References

[35] Finer TJ, Simmons RM, Spudich JA (1994). Single myosin molecule mechanics: piconewton forces and nanometre steps. *Nature* 368:113–119.

[36] Floreano D, Zufferey J-C, Srinivasan M, Ellington C (2009). *Flying Insects and Robots*. Springer-Verlag.

[37] Fukui Y (2002). Mechanistics of amoeboid locomotion: signal to forces. *Cell Biol Int* 26(11):933–44.

[38] Fukui Y, Kitanishi-Yumura T, Yumura S (1999). Myosin II-independent F-actin flow contributes to cell locomotion in dictyostelium. *J Cell Sci* 112(6):877–86.

[39] Gandhi MV, Thompson BS (2000). *Smart Materials and Structures*, Chapman & Hall.

[40] Garcia N, Stoll E (1984). Monte Carlo Simulation of Electromagnetic-Wave Scattering from Random Rough Surfaces. *Physical Review Letters* 52(20):1798–1801.

[41] Geiger B, Bershadsky A, Pankov R et al. (2001). Transmembrane crosstalk between the extracellular matrix-cytoskeleton crosstalk. *Nat Rev Mol Cell Biol* 2(11):793–805.

[42] Gittes F, Mickey B, Nettleton J et al. (1993). Flexural rigidity of microtubules and actin filaments measured from thermal fluctuations in shape. *J Cell Biol.* 120(4):923–34.

[43] Giurgiutiu V, Pomirleanu R, Rogers CA (2000). Energy-Based Comparison of Solid-State Actuators, *University of South Carolina Report*.

[44] Gupton SL, Waterman-Storer CM (2006). Spatiotemporal feedback between actomyosin and focal-adhesion systems optimizes rapid cell migration. *Cell* 125(7):1361–74.

[45] Haimo LT, Rosenbaum JL (1981). Cilia, flagella, and microtubules. *J. Cell Biol.* 91(3):125–130.

[46] Helfand BT, Chang L, Goldman RD (2004). Intermediate filaments are dynamic and motile elements of cellular architecture. *J Cell Sci* 117(2):133–41.

[47] Henson JH, Svitkina TM, Burns AR et al. (1999). Two components of actin-based retrograde flow in sea urchin coelomocytes. *Mol Biol Cell* 10(12):4075–90.

[48] Humphrey D, Duggan C, Saha D et al. (2002). Active fluidization of polymer networks through molecular motors. *Nature* 416(6879):413–6.

[49] Jay DG (2000). The clutch hypothesis revisited: ascribing the roles of actin-associated proteins in filopodial protrusion in the nerve growth cone. *J Neurobiol* 44(2):114–25.

[50] Jay PY, Pham PA, Wong SA et al. (1995). A mechanical function of myosin II in cell motility. *J Cell Sci* 108(1):387–93.

[51] Jurado C, Haserick JR, Lee J (2005). Slipping or gripping? Fluorescent speckle microscopy in fish keratocytes reveals two different mechanisms for generating a retrograde flow of actin. *Mol Biol Cell* 16(2):507–18.

[52] Kaverina I, Krylyshkina O, Small JV (2002). Regulation of substrate adhesion dynamics during cell motility. *Int J Biochem Cell Biol* 34(7):746–61.

[53] Kim S, Spenko M, Trujillo S, Heyneman B, Santos D, Cutkosky M (2008). Smooth vertical surface climbing with directional adhesion. *IEEE Transactions on Robotics* 24(1):65–74.

[54] Kolega J, Taylor DL (1993). Gradients in the concentration and assembly of myosin II in living fibroblasts during locomotion and fiber transport. *Mol Biol Cell* 4:819–836.

[55] Kruse K, Joanny JF, Jülicher F et al. (2006). Contractility and retrograde flow in lamellipodium motion. *Phys Biol* 3(2):130-7.

[56] Lee J, Leonard M, Oliver T et al. (1994). Traction forces generated by locomoting keratocytes. *J Cell Biol* 127(6 Pt 2):1957-64.

[57] Lilienthal O (1889). *Der Vogelflug als Grundlage der Fliegekunst*.

[58] Lin CH, Espreafico EM, Mooseker MS et al. (1996). Myosin drives retrograde F-actin flow in neuronal growth cones. *Neuron* 16(4):769-82.

[59] Lin CH, Forscher P (1995). Growth cone advance is inversely proportional to retrograde F-actin flow. *Neuron* 14(4):763-71.

[60] Macnab RM (1999). The bacterial flagellum: reversible rotary propellor and type III export apparatus. *J. Bacteriol.* 181(23):7149-53.

[61] McGrath JL, Eungdamrong NJ, Fisher CI et al. (2003). The force-velocity relationship for the actin-based motility of *Listeria monocytogenes*. *Curr Biol* 13(4):329-32.

[62] Medeiros NA, Burnette DT, Forscher P (2006). Myosin II functions in actin-bundle turnover in neuronal growth cones. *Nat Cell Biol* 8(3):215-26.

[63] Merz A, So M, Sheetz M (2000). Pilus retraction powers bacterial twitching motility. *Nature* 407(6800):98-102.

[64] Mitchison T, Kirschner M (1988). Cytoskeletal dynamics and nerve growth. *Neuron* 1(9):761-72.

[65] Mogilner A, Edelstein-Keshet L (2002). Regulation of actin dynamics in rapidly moving cells: a quantitative analysis. *Biophys J* 83(3):1237-58.

[66] Mogilner A, Oster G (1996). Cell motility driven by actin polymerization. *Biophys. J* 71:3030-3045.

[67] Mogilner A, Oster G (2003). Force generation by actin polymerization II: the elastic ratchet and tethered filaments. *Biophys. J* 84:1591-1605.

[68] Mogilner A, Oster G (2003). Polymer motors: pushing out the front and pulling up the back. *Curr Biol* 13(18):R721-33.

[69] Morse D (1998). Viscoelasticity of concentrated isotropic solutions of semi-flexible polymers: 1. model and stress tensor, 2. linear response. *Macromolecules* 31:7030-7044.

[70] Munevar S, Wang Y, Dembo M (2001). Traction force microscopy of migrating normal and H-ras transformed 3T3 fibroblasts. *Biophys J* 80(4):1744-57.

[71] Munevar S, Wang YL, Dembo M (2001). Distinct roles of frontal and rear cell-substrate adhesions in fibroblast migration. *Mol Biol Cell.* 12(12):3947-54.

[72] Nocks L (2007). *The robot : the life story of a technology*. Westport, CT: Greenwood Publishing Group.

[73] Oliver T, Dembo M, Jacobson K (1995). Traction forces in locomoting cells. *Cell Motil Cytoskeleton* 31(3):225-40.

[74] Pampaloni F, Lattanzi G, Jonas A et al. (2006). Thermal fluctuations of grafted microtubules provide evidence of a length-dependent persistence length. *Proc Natl Acad Sci* 103(27):10248-53.

[75] Pantaloni D, Le Clainche C, Carlier MF (2001). Mechanism of actin-based motility. *Science* 25;292(5521):1502–6.

[76] Parekh SH, Chaudhuri O, Theriot JA et al. (2005). Loading history determines the velocity of actin-network growth. *Nat Cell Biol.* 7(12):1219–23.

[77] Pollard TD, Blanchoin L, Mullins RD (2000). Molecular mechanisms controlling actin filament dynamics in nonmuscle cells. *Annu Rev Biophys Biomol Struct* 29:545–76.

[78] Pollard TD, Borisy GG (2003). Cellular motility driven by assembly and disassembly of actin filaments. *Cell* 112:453–465.

[79] Prehoda KE, Scott JA, Mullins RD, Lim WA (2000). Integration of multiple signals through cooperative regulation of the N-WASP-Arp2/3 complex. *Science* 290(5492):801–6.

[80] Raucher D, Sheetz MP (2000). Cell spreading and lamellipodial extension rate is regulated by membrane tension. *J Cell Biol* 148(1):127–36.

[81] Ridley AJ (2001). Rho family proteins: coordinating cell responses. *Trends Cell Biol* 11(12):471–7.

[82] Ridley AJ (2001). Rho GTPases and cell migration. *J Cell Sci* 114(15):2713–22.

[83] Ridley AJ, Schwartz MA, Burridge K et al. (2003). Cell migration: integrating signals from front to back. *Science* 302(5651):1704–9.

[84] Röntgen WC (1880). About the changes in shape and volume of dielectrics caused by electricity, Section III in G. Wiedemann (Ed.), *Annual Physics and Chemistry Series* 11, John Ambrosius Barth Publisher.

[85] Santoro MM, Gaudino G (2005). Cellular and molecular facets of keratinocyte reepithelization during wound healing. *Exp Cell Res* 304(1):274–286.

[86] Santos D, Heyneman B, Kim S, Esparza N, Cutkosky M (2008). Gecko-inspired climbing behaviors on vertical and overhanging surfaces. *IEEE International Conference on Robotics and Automation*, pp. 1125–1131, 2008.

[87] Schmitt OH (1957). *Emerging Science of Biophysics.* p 2.

[88] Schreiber DI, Barocas VH, Tranquilly RT (2003). Temporal variations in cell migration and traction during fibroblast-mediated gel compaction. *Biophys J* 84(6):4102–14.

[89] Spenko M, Haynes G, Saunders J, Cutkosky M, Rizzi A, Koditschek D. et al. (2008). Biologically inspired climbing with a hexapedal robot.

[90] Steltz E, Mozeika A, Rembisz J, Corson N, Jaeger HM (2010). Jamming as an enabling technology for soft robotics. *Proc. SPIE* 7642, Electroactive Polymer Actuators and Devices (EAPAD) 2010, 764225.

[91] Steltz E, Mozeika A, Rodenberg N, Brown E, Jaeger H (2009). JSEL: Jamming skin enabled locomotion. In *IEEE International Conference on Intelligent Robots and Systems* (2009).

[92] Sugiyama Y, Hirai S (2006). Crawling and Jumping by a Deformable Robot. *International Journal of Robotics Research* 25(5–6):603–620.

[93] Svitkina TM, Verkovsky AB, McQuade KM, Borisy GG (1997). Analysis of the actin-myosin II system in fish epidermal keratocytes: Mechanism of cell body translocation. *J Cell Biol* 139:397–415.

References

[94] Tan J, Shen H, Saltzman WM (2001). Micron-scale positioning of features influences the rate of polymorphonuclear leukocyte migration. *Biophys J* 81(5):2569–79.

[95] Theriot JA (2000). The polymerization motor. *Traffic* 1:19-28.

[96] Thoumine O, Kocian P, Kottelat A et al. (2000). Short-term binding of fibroblasts to fibronectin: optical tweezers experiments and probabilistic analysis. *Eur Biophys J* 29(6):398–408.

[97] Trappe V et al. (2001). Jamming phase diagram for attractive particles. *Nature* 411(6839): 772–775.

[98] Trujillo S, Heyneman B, Cutkosky M (2010). Constrained convergent gait regulation for a climbing robot. *IEEE International Conference on Robotics and Automation*, pp. 5243–5249, 2010.

[99] Unver O, Uneri A, Aydemir A, Sitti M (2006). Geckobot: a gecko inspired climbing robot using elastomer adhesives. *International Conference on Robotics and Automation*, 2329–2335.

[100] Upadhyaya A, van Oudenaarden A (2003). Biomimetic systems for studying actin-based motility. *Curr Biol* 13(18):R734–44.

[101] Vallotton P, Danuser G, Bohnet S et al. (2005). Tracking retrograde flow in keratocytes: news from the front. *Mol Biol Cell* 16(3):1223–31.

[102] Verkhovsky AB, Svitkina TM, Borisy GG (1999). Self-polarization and directional motility of cytoplasm. *Curr Biol* 9(1):11–20.

[103] Vignjevic D, Kojima S, Aratyn Y et al. (2006). Role of fascin in filopodial protrusion. *J Cell Biol* 174(6):863–75.

[104] Vincent JFV, Bogatyreva OA, Bogatyrev NR, Bowyer A, Pahl AK (2006). Biomimetics: its practice and theory. *Journal of The Royal Society* Interface 3(9):471–482.

[105] Wang FS, Liu CW, Diefenbach TJ et al. (2003). Modeling the role of myosin 1c in neuronal growth cone turning. *Biophys J* 85(5):3319–28.

[106] Watanabe N, Mitchison TJ (2002). Single-molecule speckle analysis of actin filament turnover in lamellipodia. *Science* 295(5557):1083–6.

[107] Wiesner S, Helfer E, Didry D et al. (2003). A biomimetic motility assay provides insight into the mechanism of actin-based motility. *J Cell Biol* 160(3):387–98.

[108] Yaeger JR (1984). A practical shape-memory electromechanical actuator, *Mechanical Engineering* 106:52–55.

[109] Zackroff RV, Goldman RD (1979). In vitro assembly of intermediate filaments from baby hamster kidney (BHK-21) cells. *Proc Natl Acad Sci* 76(12):6226–6230.

[110] Zaman MH, Kamm RD, Matsudaira P et al. (2005). Computational model for cell migration in three-dimensional matrices. *Biophys J* 89(2):1389–97.

[111] Zheng JQ, Wan JJ, Poo MM (1996). Essential role of filopodia in chemotropic turning of nerve growth cone induced by a glutamate gradient. *J Neurosci* 16(3):1140–9.

[112] Zufferey J-C (2008). *Bio-inspired Flying Robots: Experimental Synthesis of Autonomous Indoor Flyers*. EPFL/CRC Press.

[113] Bionics Symposium – Living Prototypes – the key to new technology, 13–15 September 1960, *Wadd Technical Report* 60-600.
[114] http://en.wikipedia.org/wiki/Biomimetics (accessed 2012).
[115] http://en.wikipedia.org/wiki/Bionics (accessed 2012).
[116] http://en.wikipedia.org/wiki/Buckminster_Fuller (accessed 2012).
[117] http://en.wikipedia.org/wiki/Codex_on_the_flight_of_birds (accessed 2012).
[118] http://en.wikipedia.org/wiki/Georges_de_Mestral (accessed 2012).
[119] http://en.wikipedia.org/wiki/Isaac_Asimov (accessed 2012).
[120] http://en.wikipedia.org/wiki/Jack_Steele (accessed 2012).
[121] http://en.wikipedia.org/wiki/Karel_%C4%8Capek (accessed 2012).
[122] http://en.wikipedia.org/wiki/Wright_brothers (accessed 2012).
[123] http://www.planetarium-jena.de/Geschichte.43.0.html (accessed 2012).
[124] http://www.saesgetters.com/sites/default/files/SmartFlex%20Compression%20&%20Tensile%20springs%20datasheets_0.pdf (accessed 2013).

I want morebooks!

Buy your books fast and straightforward online - at one of the world's fastest growing online book stores! Environmentally sound due to Print-on-Demand technologies.

Buy your books online at
www.get-morebooks.com

Kaufen Sie Ihre Bücher schnell und unkompliziert online – auf einer der am schnellsten wachsenden Buchhandelsplattformen weltweit!
Dank Print-On-Demand umwelt- und ressourcenschonend produziert.

Bücher schneller online kaufen
www.morebooks.de

OmniScriptum Marketing DEU GmbH
Heinrich-Böcking-Str. 6-8
D - 66121 Saarbrücken
Telefax: +49 681 93 81 567-9

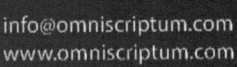
info@omniscriptum.com
www.omniscriptum.com

Printed by Books on Demand GmbH, Norderstedt / Germany